Tucholsky   Wagner   Zola   Scott   Sydow   Schlegel
   Turgenev   Wallace   Fonatne   Freud
        Twain   Walther von der Vogelweide Fouqué   Friedrich II. von Preußen
                      Weber   Freiligrath   Frey
                                    Kant   Ernst
  Fechner Fichte Weiße Rose von Fallersleben   Richthofen   Frommel
               Engels   Fielding   Hölderlin
        Fehrs   Flaubert Eichendorff Tacitus   Dumas
                  Faber   Eliasberg   Ebner Eschenbach
         Maximilian I. von Habsburg Fock   Zweig
   Feuerbach   Ewald   Eliot   Vergil
           Goethe   Elisabeth von Österreich   London
 Mendelssohn Balzac   Shakespeare   Dostojewski   Ganghofer
                   Lichtenberg   Rathenau   Doyle   Gjellerup
       Trackl   Stevenson   Hambruch
  Mommsen   Tolstoi   Lenz   Droste-Hülshoff
          Thoma   Hanrieder
  Dach   Verne   von Arnim   Hägele   Hauff   Humboldt
        Reuter   Rousseau   Hagen   Hauptmann   Gautier
      Karrillon   Garschin   Baudelaire
          Damaschke   Defoe   Hebbel
                            Descartes   Hegel   Kussmaul   Herder
  Wolfram von Eschenbach   Dickens   Schopenhauer   Rilke   George
              Darwin   Melville   Grimm Jerome   Bebel   Proust
        Bronner   Federer
            Campe   Horváth   Aristoteles   Voltaire   Herodot
  Bismarck   Vigny   Barlach   Heine
                     Gengenbach
       Storm   Casanova   Tersteegen   Gilm   Grillparzer Georgy
                 Chamberlain Lessing   Langbein   Gryphius
   Brentano   Claudius   Schiller   Lafontaine
      Strachwitz   Kralik   Iffland Sokrates
              Katharina II. von Rußland Bellamy Schilling
                                Gerstäcker   Raabe   Gibbon   Tschechow
    Löns   Hesse   Hoffmann   Gogol   Wilde   Gleim   Vulpius
         Luther   Heym   Hofmannsthal   Klee Hölty Morgenstern   Goedicke
            Roth   Heyse   Klopstock   Kleist
      Luxemburg   Puschkin   Homer
                 La Roche   Horaz   Mörike   Musil
         Machiavelli   Kierkegaard   Kraft
    Navarra Aurel   Musset   Kraus
                           Lamprecht   Kind   Hugo   Moltke
        Nestroy   Marie de France   Kirchhoff
          Nietzsche   Nansen   Laotse   Ipsen   Liebknecht
                      Marx   Lassalle   Gorki   Ringelnatz
         von Ossietzky   Klett   Leibniz
                    May   vom Stein Lawrence   Irving
          Petalozzi
                Platon   Knigge
            Sachs   Pückler   Michelangelo   Kafka
                  Poe   Kock
                                           Liebermann
                de Sade Praetorius   Mistral   Zetkin   Korolenko

The publishing house tradition has created the series **TREDITION CLASSICS**. It contains classical literature works from over two thousand years. Most of these titles have been out of print and off the bookstore shelves for decades.

The book series is intended to preserve the cultural legacy and to promote the timeless works of classical literature. As a reader of a **TREDITION CLASSICS** book, the reader supports the mission to save many of the amazing works of world literature from oblivion.

The symbol of **TREDITION CLASSICS** is Johannes Gutenberg (1400 – 1468), the inventor of movable type printing.

With the series, tradition intends to make thousands of international literature classics available in printed format again – worldwide.

All books are available at book retailers worldwide in paperback and in hardcover. For more information please visit: www.tredition.com

tradition was established in 2006 by Sandra Latusseck and Soenke Schulz. Based in Hamburg, Germany, tradition offers publishing solutions to authors and publishing houses, combined with worldwide distribution of printed and digital book content. tradition is uniquely positioned to enable authors and publishing houses to create books on their own terms and without conventional manufacturing risks.

For more information please visit: www.tredition.com

Are the Effects of Use and Disuse Inherited? An Examination of the View Held by Spencer and Darwin

W. P. (William Platt) Ball

Imprint

This book is part of the TREDITION CLASSICS series.

Author: W. P. (William Platt) Ball
Cover design: toepferschumann, Berlin (Germany)

Publisher: tredition GmbH, Hamburg (Germany)
ISBN: 978-3-8491-6661-8

www.tredition.com
www.tredition.de

Copyright:
The content of this book is sourced from the public domain.

The intention of the TREDITION CLASSICS series is to make world literature in the public domain available in printed format. Literary enthusiasts and organizations worldwide have scanned and digitally edited the original texts. tredition has subsequently formatted and redesigned the content into a modern reading layout. Therefore, we cannot guarantee the exact reproduction of the original format of a particular historic edition. Please also note that no modifications have been made to the spelling, therefore it may differ from the orthography used today.

PREFACE.

My warmest thanks are due to Mr. Francis Darwin, to Mr. E. B. Poulton (whose interest in the subject here discussed is shown by his share in the translation of Weismann's *Essays on Heredity*), and to Professor Romanes, for the help afforded by their kindly suggestions and criticisms, and for the advice and recommendation under which this essay is now published. Encouragement from Mr. Francis Darwin is to me the more precious, and the more worthy of grateful recognition, from the fact that my general conclusion that acquired characters are *not* inherited is [vi] at variance with the opinion of his revered father, who aided his great theory by the retention of some remains of Lamarck's doctrine of the inherited effect of habit. I feel as if the son, as representative of his great progenitor, were carrying out the idea of an appreciative editor who writes to me: "We must say that if Darwin were still alive, he would find your arguments of great weight, and undoubtedly would give to them the serious consideration which they deserve." I hope, then, that I may be acquitted of undue presumption in opposing a view sanctioned by the author of the *Origin of Species*, but already stoutly questioned and firmly rejected by such followers of his as Weismann, Wallace, Poulton, Ray Lankester, and others, to say nothing of its practical rejection by so great an authority on heredity as Francis Galton.

The sociological importance of the subject has already been insisted on in emphatic terms by [vii] Mr. Herbert Spencer, and this importance may be even greater than he imagined.

Civilization largely sets aside the harsh but ultimately salutary action of the great law of Natural Selection without providing an efficient substitute for preventing degeneracy. The substitute on which moralists and legislators rely—if they think on the matter at all—is the cumulative inheritance of the beneficial effects of education, training, habits, institutions, and so forth—the inheritance, in short, of acquired characters, or of the effects of use and disuse. If this substitute is but a broken reed, then the deeper thinkers who gradually teach the teachers of the people, and ultimately even influence the legislators and moralists, must found their systems of morality and their criticisms of social and political laws and institutions and customs and ideas on the basis of the Darwinian law rather than on that of Lamarck. [viii]

Looking forward to the hope that the human race may become consciously and increasingly master of itself and of its destiny, and recognizing the Darwinian principle of the selection of the fittest as the *only* means of preventing the moral and physical degeneracy which, like an internal dry rot, has hitherto been the besetting danger of all civilizations, I desire that the thinkers who mould the opinions of mankind shall not be led astray from the true path of enduring progress and happiness by reliance on fallacious beliefs which will not bear examination. Such, at least, is the feeling or motive which has prompted me to devote much time and thought to a difficult but important inquiry in a debatable region of inference and conjecture, where (I am afraid) evidence on either side can never be absolutely conclusive, and where, especially, the absolute demonstration of a universal negative cannot reasonably be expected.

[ix]

CONTENTS.

PREFACE
IMPORTANCE AND BEARING OF THE INQUIRY
SPENCER'S EXAMPLES AND ARGUMENTS
Diminution of the Jaws
Diminished Biting Muscles of Lap-Dogs
Crowded Teeth
Blind Cave-Crabs
No Concomitant Variation from Concomitant Disuse
The Giraffe, and Necessity for Concomitant Variation
Alleged Ruinous Effects of Natural Selection
Adverse Case of Neuter Insects
Æsthetic Faculties
Lack of Evidence
Inherited Epilepsy in Guinea-Pigs
Inherited Insanity and Nervous Disorders
Individual and Transmissible Type not Modified Alike
DARWIN'S EXAMPLES
Reduced Wings of Birds of Oceanic Islands
Drooping Ears and Deteriorated Instincts
Wings and Legs of Ducks and Fowls
Pigeons' Wings
Shortened Breast-Bone in Pigeons
Shortened Feet in Pigeons
Shortened Legs of Rabbits
Blind Cave-Animals
Inherited Habits
Tameness of Rabbits
Modifications Obviously Attributable to Selection
Similar Effects of Natural Selection and Use-Inheritance
Inferiority of Senses in Europeans
Short-Sight in Watchmakers and Engravers
Larger Hands of Labourers' Infants

Thickened Sole in Infants
A Source of Mental Confusion
Weakness of Use-Inheritance
INHERITED INJURIES
Inherited Mutilations
The Motmot's Tail
Other Inherited Injuries Mentioned by Darwin
Quasi-Inheritance
MISCELLANEOUS CONSIDERATIONS
True Relation of Parents and Offspring
Inverse Inheritance
Early Origin of the Ova
Marked Effects of Use and Disuse on the Individual
Would Natural Selection Favour Use-Inheritance?
Use-Inheritance an Evil
Varied Effects of Use and Disuse
Use-Inheritance Implies Pangenesis
Pangenesis Improbable
Spencer's Explanation of Use-Inheritance
CONCLUSIONS
Use-Inheritance Discredited as Unnecessary, Unproven, and Improbable
Modern Reliance on Use-Inheritance Misplaced

[1]

ARE THE EFFECTS OF USE AND DISUSE INHERITED?

IMPORTANCE AND BEARING OF THE INQUIRY.

The question whether the effects of use and disuse are inherited, or, in other words, whether acquired characters are hereditary, is of considerable interest to the general student of evolution; but it is, or should be, a matter of far deeper interest to the thoughtful philanthropist who desires to ensure the permanent welfare and happiness of the human race. So [2] profoundly important, in fact, are the moral, social, and political conclusions that depend on the answer to this inquiry, that, as Mr. Herbert Spencer rightly says, it "demands, beyond all other questions whatsoever, the attention of scientific men."

It is obvious that we can produce important changes in the individual. We can, for example, improve his muscles by athletics, and his brain by education. The use of organs enlarges and strengthens them; the disuse of parts or faculties weakens them. And so great is the power of habit that it is proverbially spoken of as "second nature." It is thus certain that we can modify the individual. We can strengthen (or weaken) his body; we can improve (or deteriorate) his intellect, his habits, his morals. But there remains the still more important question which we are about to consider. Will such modifications be inherited by the offspring of the [3] modified individual? Does individual improvement transmit itself to descendants independently of personal teaching and example? Have artificially produced changes of structure or habit any inherent tendency to become congenitally transmissible and to be converted in time into fixed traits of constitution or character? Can the philanthropist rely on such a tendency as a hopeful factor in the evolution of mankind? — the only sound and stable basis of a higher and happier state of things being, as he knows or ought to know, the innate and constitutionally-fixed improvement of the race as a whole. If acquired modifications are impressed on the offspring and on the race, the systematic moral training of individuals will in time produce a constitutionally moral race, and we may hope to improve mankind even in defiance of the unnatural selection by which a spurious but highly popular philanthropy would systematically favour the survival of the unfittest [4] and the rapid multiplication of the worst. But if acquired modifications do not tend to be transmitted, if the use or disuse of organs or faculties does not similarly affect posterity by inheritance, then it is evident that no innate improvement in the race can take place without the aid of natural or artificial selection.

Herbert Spencer maintains that the effects of use and disuse *are* inherited in kind, and in his *Factors of Organic Evolution* [1] he has supported his contention with a selection of facts and reasonings which I shall have the temerity to examine and criticize. Darwin also held the same view, though not so strongly. And here, to prevent misunderstanding, I may say that the admiration and reverence and gratitude due to Darwin ought not to be allowed to interfere in the slightest [5] degree with the freest criticism of his conclusions. To perfect his work by the correction of really extraneous errors is as much a sacred duty as to study and apply the great truths he has taught.

FOOTNOTES:

[1] Which originally appeared in the *Nineteenth Century* for April and May, 1886.

[6]

SPENCER'S EXAMPLES AND ARGUMENTS.

DIMINUTION OF THE JAWS IN CIVILIZED RACES.

Mr. Spencer verified this by comparing English jaws with Australian and Negro jaws at the College of Surgeons. [2] He maintains that the diminution of the jaw in civilized races can *only* have been [7] brought about by inheritance of the effects of lessened use. But if English jaws are lighter and thinner than those of Australians and Negroes, so too is the rest of the skull. As the diminution in the weight and thickness of the walls of the cranium cannot well be ascribed to disuse, it must be attributed to some other cause; and this cause may have affected the jaw also. Cessation of the process by which natural selection [3] favoured strong thick bones during ages of brutal violence might bring about a change in this direction. Lightness of structure, facilitating agility and being economical of material, would also be favoured by natural selection so far as strength was not too seriously diminished.

[8]

Sexual selection powerfully affects the human face, and so must affect the jaws—as is shown by the differences between male and female jaws, and by the relative lightness and smallness of the latter, especially in the higher races. Human preference, both sexual and social, would tend to eliminate huge jaws and ferocious teeth when these were no longer needed as weapons of war or organs of prehension, &c. We can hardly assume that the lower half of the face is specially exempt from the influence of natural and sexual selection; and the effects of these undoubted factors of evolution must be fully considered before we are entitled to call in the aid of a factor whose existence is questioned.

After allowing for lost teeth and the consequent alveolar absorption, and for a reduction proportional to that shown in the rest of the skull, the difference in average weight in fifty European and fourteen Australian male jaws [9] at the College of Surgeons turned out to be less than a fifth of an ounce, or about 5 per cent. This slight reduction may be much more than accounted for by such causes as disuse in the individual, human preference setting back the teeth, and partial transference of the much more marked diminution seen in female jaws. There is apparently no room for accumulated *inherited* effects of ancestral disuse. The number of jaws is small, indeed; but weighing them is at least more decisive than Mr. Spencer's mere inspection.

The differences between Anglo-Saxon male jaws and Australian and Tasmanian jaws are most easily explained as effects of human preference and natural selection. We can hardly suppose that disuse would maintain or develop the projecting chin, increase its perpendicular height till the jaw is deepest and strongest at its extremity, evolve a side flange, and enlarge [10] the upper jaw-bone to form part of a more prominent nose, while drawing back the savagely obtrusive teeth and lips to a more pleasing and subdued position of retirement and of humanized beauty. If human preference and natural selection caused some of these differences, why are they incompetent to effect changes in the direction of a diminution of the jaw or teeth? And if use and disuse are the sole modifying agents in the case of the human jaw, why should men have any more chin than a gorilla or a dog?

The excessive weight of the West African jaws at the College of Surgeons is partly *against* Mr. Spencer's contention, unless he assumes that Guinea Negroes use their jaws far more than the Australians, a supposition which seems extremely improbable. The heavier skull and narrower molar teeth point however to other factors than increased use. [11]

The striking variability of the human jaw is strongly opposed to the idea of its being under the direct and dominant control of so uniform a cause as ancestral use and disuse. Mr. Spencer regards a variation of 1 oz. as a large one, but I found that the English jaws in the College of Surgeons varied from 1·9 oz. to 4·3 oz. (or 5 oz. if lost teeth were allowed for); Australian jaws varied from 2 oz. to 4·5 oz. (with *no* lost teeth to allow for); while in Negro jaws the maximum rose to over 5½ oz. [4] In spite of disuse some European jaws were twice as heavy as the lightest Australian jaw, either absolutely or (in some cases) relatively to the cranium. The uniformity of change relied upon by Mr. Spencer is scarcely borne out by the facts so far as male jaws are concerned. The great reduction in the weight of *female* jaws *and skulls* evidently points to [12] sexual selection and to panmixia under male protection.

I think, on the whole, we must conclude that the human jaws do not afford satisfactory proof of the inheritance of the effects of use and disuse, inasmuch as the differences in their weight and shape and size can be more reasonably and consistently accounted for as the result of less disputable causes.

DIMINISHED BITING MUSCLES OF LAP-DOGS.

The next example, the reduced biting muscles, &c., of lap-dogs is also unsatisfactory as a proof of the inheritance of the effects of disuse; for the change can readily be accounted for without the introduction of such a factor. The previous natural selection of strong jaws and teeth and muscles is reversed. The conscious or unconscious selection of lap-dogs with the least tendency to [13] bite would easily bring about a general enfeeblement of the whole biting apparatus—weakness of the parts concerned favouring harmlessness. Mr. Spencer maintains that the dwindling of the parts concerned in clenching the jaw is certainly not due to artificial selection because the modifications offer no appreciable external signs. Surely hard biting is sufficiently appreciable by the person bitten without any visual admeasurement of the masseter muscles or the zygomatic arches. Disuse during lifetime would also cause some amount of degeneracy; and I am not sure that Mr. Spencer is right in *entirely* excluding economy of nutrition from the problem. Breeders would not over-feed these dogs; and the puppies that grew most rapidly would usually be favoured.

[14]

CROWDED TEETH.

The too closely-packed teeth in the "decreasing" jaws of modern men (p. 13) [5] are also suggestive of other causes than use and disuse. Why is there not simultaneous variation in teeth and jaws, if disuse is the governing factor? Are we to suppose that the size of the human teeth is maintained by use at the same time that the jaws are being diminished by disuse? Mr. Spencer acknowledges that the crowding of bull-dogs' and lap-dogs' teeth is caused by the artificial selection of shortened jaws. If a similar change is really occurring in man, could it not be similarly explained by some factor, such as sexual selection, which might affect the outward appearance at the cost of less obvious defects or inconveniences?

Mr. Spencer points to the decay of modern teeth as a sign or result of their being overcrowded [15] through the diminution of the jaw by disuse. [6] But the teeth which are the most frequently overcrowded are the lower incisors. The upper incisors are less overcrowded, being commonly pressed outwards by the lower arc of teeth fitting inside them in biting. The lower incisors are correspondingly pressed inwards and closer together. Yet the upper incisors decay—or at least are extracted—about twenty times as frequently as the closely packed lower incisors. [7] Surely this must indicate that the cause of decay is not overcrowding.

[16]

The lateness and irregularity of the wisdom teeth are sometimes supposed to indicate their gradual disappearance through want of room in a diminishing jaw. But a note on Tasmanian skulls in the *Catalogue of the College of Surgeons* (p. 199) shows that this lateness and irregularity have been common among Tasmanians as well as among civilized races, so that the change can hardly be attributed to the effects of disuse under civilization.

[17]

BLIND CAVE-CRABS.

The cave-crabs which have lost their disused eyes but *not the disused eye-stalks* appear to illustrate the effects of natural selection rather than of disuse. The loss of the exposed, sensitive, and worse-than-useless eye, would be a decided gain, while the disused eye-stalk, being no particular detriment to the crab, would be but slightly affected by natural selection, though open to the cumulative effects of disuse. The disused but better protected eyes of the blind cave-rat are still "of large size" (*Origin of Species*, p. 110).

NO CONCOMITANT VARIATION FROM CONCOMITANT DISUSE.

It is but fair to add that these instances of the cave-crab's eye-stalk and the closely-packed teeth are put forward by Mr. Spencer with the more [18] immediate object of proving that there is "no concomitant variation in co-operative parts," even when "formed out of the same tissue, like the crab's eye and its peduncle" (pp. 12-14, 23, 33). It escapes his notice, however, that in two out of his three cases it is *disuse*, or *diminished use*, which fails to cause concomitant variation or proportionate variation.

THE GIRAFFE, AND NECESSITY FOR CONCOMITANT VARIATION.

Having unwittingly shown that lessened use of closely-connected and co-operative parts does not cause concomitant variation in these parts, Mr. Spencer concludes that the concomitant variation requisite for evolution can only be caused by altered degrees of use or disuse. He elaborately argues that the many co-ordinated modifications of parts necessitated by each important [19] alteration in an animal are so complex that they cannot possibly be brought about except by the inherited effect of the use and disuse of the various parts concerned. He holds, for instance, that natural selection is inadequate to effect the numerous concomitant changes necessitated by such developments as that of the long neck of the giraffe. Darwin, however, on the contrary, holds that natural selection alone "would have sufficed for the production of this remarkable quadruped." [8] He is surprised at Mr. Spencer's view that natural selection can do so little in modifying the higher animals. Thus one of the chief arguments with which Mr. Spencer supports his theory is so poorly founded as to be rejected by a far greater authority on such subjects. All that is needed is that natural selection should preserve the tallest giraffes through times of famine by their [20] being able to reach otherwise inaccessible stores of foliage. The continual variability of all parts of the higher animals gives scope for innumerable changes, and Nature is not in a hurry. Mr. Spencer, however, says that "the chances against any adequate readjustments fortuitously arising must be infinity to one." But he has also shown that altered degree of use does not cause the needed concomitant variation of co-operative parts. So the chances against a beneficial change in an animal must be, at a liberal estimate, infinity to two. Mr. Spencer, if he has proved anything, has proved that it is practically impossible that the giraffe can have acquired a long neck, or the elk its huge horns, or that any species has ever acquired any important modification.

Mr. Wallace, in his *Darwinism*, answers Mr. Spencer by a collection of facts showing that "variation is the rule," that the range of variation in wild animals and plants is much greater [21] than was supposed, and that "each part varies to a considerable extent independently" of other parts, so that "the materials constantly ready for natural selection to act upon are abundant in quantity and very varied in kind." While co-operative parts would often be more or less correlated, so that they would tend to vary together, coincident variation is not necessary. The lengthened wing might be gained in one generation, and the strengthened muscle at a subsequent period; the bird in the meanwhile drawing upon its surplus energy, aided (as I would suggest) by the strengthening effect of increased use in the individual. Seeing that artificial selection of complicated variations has modified animals in many points either simultaneously or by slow steps, as with otter-sheep, fancy pigeons, &c. (many of the characters thus obtained being clearly independent of use and disuse), natural selection must be credited with similar powers, [22] and Mr. Wallace concludes that Mr. Spencer's insuperable difficulty is "wholly imaginary."

The extract concerning a somewhat similar "class of difficulties," which Mr. Spencer quotes from his *Principles of Biology*, is faulty in its reasoning, [9] though legitimate in its conclusion concerning the increasing difficulty of evolution in proportion with the increasing number and complexity of faculties to be evolved. But this increasing difficulty of complex evolution is only overcome by *some* favourably-varying individuals and species—not by all. And as the difficulty [23] increases we find neglect and decay of the less-needed faculties—as with domesticated animals and civilized men, who lose in one direction while they gain in another. The increasing difficulty of complex evolution by natural selection is no proof whatever of use-inheritance [10] except to those who confound difficulty with impossibility.

ALLEGED RUINOUS EFFECTS OF NATURAL SELECTION.

Mr. Spencer further contends that natural selection, by unduly developing specially advantageous modifications without the necessary but complex secondary modifications, would render the constitution [24] of a variety "unworkable" (p. 23). But this seems hardly feasible, seeing that natural selection must continually favour the most workable constitutions, and will only preserve organisms in proportion as they combine general workableness with the special modification. On the other hand, according to Mr. Spencer himself, use-inheritance must often disturb the balance of the constitution. Thus it tends to make the jaws and teeth unworkable through the overcrowding and decay of the teeth—there being, as his illustrations show, no simultaneous or concomitant or proportional variation in relation to altered degree of use or disuse.

ADVERSE CASE OF NEUTER INSECTS.

Mr. Spencer also holds that most mental phenomena, especially where complex or social or moral, can only be explained as arising from [25] use-inheritance, which becomes more and more important as a factor of evolution as we advance from the vegetable world and the lower grades of animal life to the more complex activities, tastes, and habits of the higher organizations (preface, and p. 74). But there happens to be a tolerably clear proof that such changes as the evolution of complicated structures and habits and social instincts *can* take place independently of use-inheritance. The wonderful instincts of the working bees have apparently been evolved (at least in all their later social complications and developments) without the aid of use-inheritance—nay, in spite of its utmost opposition. Working bees, being infertile "neuters," cannot as a rule transmit their own modifications and habits. They are descended from countless generations of queen bees and drones, whose habits have been widely different from those of the workers, and whose structures are dissimilar in [26] various respects. In many species of ants there are two, and in the leaf-cutting ants of Brazil there are *three*, kinds of neuters which differ from each other and from their male and female ancestors "to an almost incredible degree." [11] The soldier caste is distinguished from the workers by enormously large heads, very powerful mandibles, and "extraordinarily different" instincts. In the driver ant of West Africa one kind of neuter is three times the size of the other, [27] and has jaws nearly five times as long. In another case "the workers of one caste alone carry a wonderful sort of shield on their heads." One of the three neuter classes in the leaf-cutting ants has a single eye in the midst of its forehead. In certain Mexican and Australian ants some of the neuters have huge spherical abdomens, which serve as living reservoirs of honey for the use of the community. In the equally wonderful case of the termites, or so-called "white ants" (which belong, however, to an entirely different order of insect from the ants and bees) the neuters are blind and wingless, and are divided into soldiers and workers, each class possessing the requisite instincts and structures adapting it for its tasks. Seeing that natural selection can form and maintain the various structures and the exceedingly complicated instincts of ants and bees and wasps and termites in direct defiance of the alleged tendency to use-inheritance, surely [28] we may believe that natural selection, unopposed by use-inheritance, is equally competent for the work of complex or social or mental evolution in the many cases where the strong presumptive evidence cannot be rendered almost indisputable by the exceptional exclusion of the modified animal from the work of reproduction.

Ants and bees seem to be capable of altering their habits and methods of action much as men do. Bees taken to Australia cease to store honey after a few years' experience of the mild winters. Whole communities of bees sometimes take to theft, and live by plundering hives, first killing the queen to create dismay among the workers. Slave ants attend devotedly to their captors, and fight against their own species. Forel reared an artificial ant-colony made up of five different and more or less hostile species. Why cannot a much more intelligent animal modify his habits far more rapidly and comprehensively [29] without the aid of a factor which is clearly unnecessary in the case of the more intelligent of the social insects?

ÆSTHETIC FACULTIES.

The modern development of music and harmony (p. 19) is undeniable, but why could it only have been brought about by the help of the inheritance of the effects of use? Why are we to suppose that "minor traits" such as the "æsthetic perceptions" cannot have been evolved by natural selection (p. 20) or by sexual selection? Darwin holds that our musical faculties were developed by sexual preference long before the acquisition of speech. He believes that the "rhythms and cadences of oratory are derived from previously developed musical powers" — a conclusion "exactly opposite" to that arrived at [30] by Mr. Spencer. [12] The emotional susceptibility to music, and the delicate perceptions needed for the higher branches of art, were apparently the work of natural and sexual selection in the long past. Civilization, with its leisure and wealth and accumulated knowledge, perfects human faculties by artificial cultivation, develops and combines means of enjoyment, and discovers unsuspected sources of interest and pleasure. The sense of harmony, modern as it seems to be, must have been a latent and indirect consequence of the development of the sense of hearing and of melody. Use, at least, could never have called it into existence. Nature favours and develops enjoyments to a certain extent, for they subserve self-preservation and sexual and social preference in innumerable ways. But modern æsthetic advance seems to be almost entirely due to the culture of latent abilities, the formation of complex [31] associations, the selection and encouragement of talent, and the wide diffusion and imitation of the accumulated products of the well-cultivated genius of favourably varying individuals. The fact that uneducated persons do not enjoy the higher tastes, and the rapidity with which such tastes are acquired or professed, ought to be sufficient proof that modern culture is brought about by far swifter and more potent influences than use-inheritance. Neither would this hypothetical factor of evolution materially aid in explaining the many other rapid changes of habit brought about by education, custom, and the changed conditions of civilization generally. Powerful tastes — as is incontestably shown in the cases of alcohol and tobacco — lie latent for ages, and suddenly become manifest when suitable conditions arise. Every discovery, and each step in social and moral evolution, produces its wide-spreading train of consequences. I see no reason why use-inheritance [32] need be credited with any share in the cumulative results of the invention of printing and the steam-engine and gunpowder, or of freedom and security under representative government, or of science and art and the partial emancipation of the mind of man from superstition, or of the innumerable other improvements or changes that take place under modern civilization.

Mr. Spencer suggests an inquiry whether the greater powers possessed by eminent musicians were not mainly due to the inherited effect of the musical practice of their fathers (p. 19). But these great musicians inherited far more than their parents possessed. The excess of their powers beyond their parents' must surely be attributed to spontaneous variation; and who shall say that the rest was in any way due to use-inheritance? If, too, the superiority of geniuses proves use-inheritance, why should not the inferiority of the sons of geniuses prove the existence of a [33] tendency which is the exact opposite of use-inheritance? But nobody collects facts concerning the degenerate branches of musical families. Only the favourably varying branches are noticed, and a general impression of rapid evolution of talent is thus produced. Such cases might be explained, too, by the facts that musical faculty is strong in both sexes, that musical families associate together, and that the more gifted members may intermarry. Great musicians are often astonishingly precocious. Meyerbeer "played brilliantly" at the age of six. Mozart played beautifully at four. Are we to suppose that the effect of the *adult* practice of parents was inherited at this early age? If use-inheritance was not necessary in the case of Handel, whose father was a surgeon, why is it needed to account for Bach?

[34]

LACK OF EVIDENCE.

The "direct proofs" of use-inheritance are not as plentiful as might be desired, it appears (pp. 24-28). This acknowledged "lack of recognized evidence" is indeed the weakest feature in the case, though Mr. Spencer would fain attribute this lack of direct proof to insufficient investigation and to the inconspicuous nature of the inheritance of the modification. But there is an almost endless abundance of conspicuous examples of the effects of use and disuse in the individual. How is it that the subsequent inheritance of these effects has not been more satisfactorily observed and investigated? Horse-breeders and others could profit by such a tendency, and one cannot help suspecting that the reason they ignore it must be its practical inefficacy, arising probably from its weakness, its obscurity and uncertainty or its non-existence.

[35]

INHERITED EPILEPSY IN GUINEA-PIGS.

Brown-Séquard's discovery that an epileptic tendency artificially produced by mutilating the nervous system of a guinea-pig is occasionally inherited may be a fact of "considerable weight," or on the other hand it may be entirely irrelevant. Cases of this kind strike one as peculiar exceptions rather than as examples of a general rule or law. They seem to show that certain morbid conditions may occasionally affect both the individual and the reproductive elements or transmissible type in a similar manner; but then we also know that such prompt and complete transmission of an artificial modification is widely different from the usual rule. Exceptional cases require exceptional explanations, and are scarcely good examples of the effect of a general tendency which in almost all other cases is so inconspicuous in its immediate effects. Further remarks on this [36] inherited epilepsy can be most conveniently introduced later on in connection with Darwin's explanation of the inherited mutilation which it usually accompanies, but which Mr. Spencer does not mention.

INHERITED INSANITY AND NERVOUS DISORDERS.

Mr. Spencer infers that, because insanity is usually hereditary, and insanity can be artificially produced by various excesses, therefore this artificially-produced insanity must also be hereditary (p. 28). Direct evidence of this conclusion would be better than a mere inference which may beg the very question at issue. That the liability to insanity commonly runs in families is no proof that strictly non-inherited insanity will subsequently become hereditary. I think that theories should be based on facts rather than facts on [37] theories, especially when those facts are to be the basis or proof of a further theory.

Mr. Spencer also points out that he finds among physicians "the belief that nervous disorders of a less severe kind are inheritable"—a general belief which does not necessarily include the transmission of purely artificially-produced disorders, and so misses the point which is really at issue. He proceeds, however, to state more definitely that "men who have prostrated their nervous systems by prolonged overwork or in some other way, have children more or less prone to nervousness." The following observations will, I think, warrant at least a suspension of judgment concerning this particular form of use-inheritance.

(1) The nervousness is seen in the *children* at an early age, although the nervous prostration from which it is supposed to be derived obviously occurs in the parent at a much later period of life. This change in time is contrary to the rule [38] of inheritance at corresponding periods; and, together with the unusual promptness and comparative completeness of the inheritance, it may indicate a special injury or deterioration of the reproductive elements rather than true inheritance. The healthy brain of early life has failed to transmit its robust condition. Is use-inheritance, then, only effective for evil? Does it only transfer the newly-acquired weakness, and not the previous long-continued vigour?

(2) Members of nervous families would be liable to suffer from nervous prostration, and by the ordinary law of heredity alone would transmit nervousness to their children.

(3) The shattered nerves or insanity resulting from alcoholic and other excesses, or from overwork or trouble, are evidently signs of a grave constitutional injury which may react upon the reproductive elements nourished and developed in that ruined constitution. The deterioration in [39] parent and child may often display itself in the same organs—those probably which are hereditarily weakest. Acquired diseases or disorders thus appear to be transmitted, when all that was conveyed to the offspring was the exciting cause of a lowered vitality or disordered action, together with the ancestral liability to such diseases under such conditions.

(4) Francis Galton says that "it is hard to find evidence of the power of the personal structure to react upon the sexual elements, that is not open to serious objection." Some of the cases of apparent inheritance he regards as coincidence of effect. Thus "the fact that a drunkard will often have imbecile children, although his offspring previous to his taking to drink were healthy," is an "instance of simultaneous action," and not of true inheritance. "The alcohol pervades his tissues, and, of course, affects the germinal matter in the sexual elements as much as it does [40] that in his own structural cells, which have led to an alteration in the quality of his own nerves. Exactly the same must occur in the case of many constitutional diseases that have been acquired by long-continued irregular habits." [13]

INDIVIDUAL AND TRANSMISSIBLE TYPE NOT MODIFIED ALIKE BY THE DIRECT EFFECT OF CHANGED HABITS OR CONDITIONS.

Mr. Spencer finds it hard to believe that the modifications conveyed to offspring are not identical in tendency with the changes effected in the parent by altered use or habit (pp. 23-25, 34). But it is perfectly certain that the two sets of effects do not necessarily correspond. The effect of changed habits or conditions on the individual is often very far from coinciding with the effects on the reproductive elements or the [41] transmissible type. The reproductive system is "extremely sensitive" to very slight changes, and is often powerfully affected by circumstances which otherwise have little effect on the individual (*Origin of Species*, p. 7). Various animals and plants become sterile when domesticated or supplied with too much nourishment. The native Tasmanians have already become extinct from sterility caused by greatly changed diet and habits. If, as Mr. Spencer teaches, continued culture and brain-work will in time produce lessened fertility or comparative sterility, we may yet have to be careful that intellectual development does not become a species of suicide, and that the culture of the race does not mean its extinction — or at least the extinction of those most susceptible of culture.

The reproductive elements are also disturbed and modified in innumerable minor ways. Changed conditions or habits tend to produce [42] a general "plasticity" of type, the "indefinite variability" thus caused being apparently irrelevant to the change, if any, in the individual. [14] A vast number of variations of structure have certainly arisen independently of similar parental modification as the preliminary. Whatever first caused these "spontaneous" congenital variations affected the reproductive elements quite differently from the individual. "When a new peculiarity first appears we can never predict whether it will be inherited." Many varieties of plants only keep true from shoots, and not from seed, which is by no means acted on in the same way as the individual plant. Seeing that such plants have [43] *two* reproductive types, both constant, it is evident that these cannot both be modified in the same way as the parent is modified. Many parental modifications of structure and habit are certainly not conveyed to neuter ants and bees; other modifications, which are not seen in the parents, being conveyed instead. Many other circumstances tend to show that the individual and the transmissible type are independent of each other so far as modifications of parts are concerned.

It may seem natural to expect the transmission of an enlarged muscle or a cultivated brain, but, on the other hand, why should it be unreasonable to expect that a modification which was non-congenital in origin should still remain non-congenital? Why should the non-transmission of that which was not transmitted be surprising?

Mr. Spencer thinks that the non-transmission of acquired modifications is incongruous with the [44] great fact of atavism. But the great law of the inheritance of that which is a development of the transmissible type does not necessarily imply the inheritance of modifications acquired by the individual. Because English children may inherit blue eyes and flaxen hair from their Anglo-Saxon ancestors, it by no means follows that an Englishman must inherit his father's sunburnt complexion or smooth-shaven face. Of course atavism ultimately adopts many instances of revolt against its sway. But to assume that these changes of type *follow* the personal change rather than cause it, is to assume the whole question at issue. That like begets like is true as a broad principle, but it has many exceptions, and the non-heredity of acquired characters may be one of them.

FOOTNOTES:

[2] *Principles of Biology*, § 166, footnote. The English jaws are somewhat lighter than the Australian jaws, though I could not undertake to affirm that they are really shorter and smaller. In the typical skulls depicted on p. 68 of the official guide to the mammalian galleries at South Kensington, the typical Caucasian jaw is very much larger than the Tasmanian jaw, although the repulsively obtrusive teeth of the latter convey the contrary idea to the imagination. Mr. Spencer's assumption that the ancient Britons had large jaws appears to me erroneous. (See Professor Rolleston's *Scientific Papers and Addresses*, i. p. 250.)

[3] Romanes, Galton, and Weismann have made great use of this principle in explaining the diminution of disused organs. Weismann has given it the name of *Panmixia*,—*all* individuals being equally free to survive and commingle their variations, and not merely selected or favoured individuals. See his *Essays on Heredity*, &c., p. 90 (Clarendon Press).

[4] Inclusive in each case of fixed strengthening wire weighing about a sixteenth of an ounce or less.

[5] References of course are to *Factors of Organic Evolution*.

[6] P. 13; and *Nineteenth Century*, February, 1888, p. 211.

[7] Tomes's *Dental Surgery*, pp. 273-275. Tomes observes that it is as yet uncertain in what way civilization predisposes to caries. But he shows that caries is caused by the lime salts in the teeth being attacked by *acids* from decomposing food in crevices, from artificial drink such as cyder, from sugar, from medicine, and from vitiated secretions of the mouth. It is evident that in civilized races natural selection cannot so rigorously insist on sound teeth, sound constitutions, and *protective alkaline* saliva. The reaction of the civilized mouth is often acid, especially when the system is disordered by dyspepsia or other diseases or forms of ill-health common under civilization. The main supply of saliva, which is poured from the cheeks opposite the upper molars, is often acid when in small quantities. But the submaxillary and sub-lingual saliva poured out at the foot of the lower incisors and held in the front part of the jaw as in a spoon, "differs from parotid saliva in being more alkaline" (Foster's

Text Book of Physiology, p. 238; Tomes, pp. 284, 685). One observer says that the reaction near the lower incisors is "never acid." Hence (I conclude) the remarkable immunity of the lower incisors and canines from decay, an immunity which extends backwards in a lessening degree to the first and second bicuspids. The close packing of the lower incisors may assist by preventing the retention of decaying fragments of food. Sexual selection may promote caries by favouring white teeth, which are more prone to decay than yellow ones. Acid vitiation of the mucus might account both for caries and (possibly) for the strange infertility of some inferior races under civilization.

[8] *Origin of Species*, pp. 198-9; *Variation of Animals and Plants under Domestication*, vol. ii. p. 328 footnote, also p. 206.

[9] Mr. Spencer weakly argues that an advantageous attribute (such as swiftness, keen sight, courage, sagacity, strength, &c.) cannot be increased by natural selection unless it is "of greater importance, for the time being, than most of the other attributes"; and that natural selection cannot develop any one superiority when animals are equally preserved by "other superiorities." But as natural selection will simultaneously eliminate tendencies to slowness, blindness, deafness, stupidity, &c., it *must* favour and improve many points simultaneously, although no one of them may be of greater importance than the rest. Of course the more complicated the evolution the slower it will be; but time is plentiful, and the amount of elimination is correspondingly vast.

[10] I venture to coin this concise term to signify *the direct inheritance of the effects of use and disuse in kind*. Having a name for a thing is highly convenient; it facilitates clearness and accuracy in reasoning, and in this particular inquiry it may save some confusion of thought from double or incomplete meanings in the shortened phrases which would otherwise have to be employed to indicate this great but nameless factor of evolution.

[11] *Origin of Species*, pp. 230-232; Bates's *Naturalist on the Amazons*. Darwin is "surprised that no one has hitherto advanced the demonstrative case of neuter insects, against the well-known doctrine of inherited habit, as advanced by Lamarck." As he justly observes, "it proves that with animals, as with plants, any amount of

modification may be effected by the accumulation of numerous, slight, spontaneous variations, which are in any way profitable, without exercise or habit having been brought into play. For peculiar habits confined to the workers or sterile females, however long they might be followed, could not possibly affect the males and fertile females, which alone leave any descendants." Some slight modification of these remarks, however, may possibly be needed to meet the case of "factitious queens," who (probably through eating particles of the royal food) become capable of producing a few male eggs.

[12] *Descent of Man*, pp. 573, 572, and footnote.

[13] *Contemporary Review*, December, 1875, p. 92.

[14] See *Origin of Species*, pp. 5-8. "Changed conditions induce an almost indefinite amount of fluctuating variability, by which the whole organization is rendered in some degree plastic" (*Descent of Man*, p. 30). It also appears that "the nature of the conditions is of subordinate importance in comparison with the nature of the organism in determining each particular form of variation;—perhaps of not more importance than the nature of the spark, by which a mass of combustible matter is ignited, has in determining the nature of the flames" (*Origin of Species*, p. 8).

[45]

DARWIN'S EXAMPLES.

The most formidable cases brought forward by Mr. Spencer are from Darwin. I shall endeavour to show, however, that Darwin was probably wrong in retaining the older explanation of these facts, and that the remains of the Lamarckian theory of use-inheritance need not any longer encumber the great explanation which has superseded that fallacious and unproven theory and has rendered it totally unnecessary. Meanwhile I think it is an excellent sign that Mr. Spencer has to complain that "Nowadays most naturalists are more Darwinian than Mr. Darwin himself"—inasmuch as they are inclined to say that there is "no proof" that the effects of use and disuse are inherited. [46] Other excellent signs are the recent issue of a translation of Weismann's important essays on this and kindred subjects, [15] the strong support given to his views by Wallace in his *Darwinism*, and their adoption by Ray Lankester in his article on Zoology in the latest edition of the *Encyclopædia Britannica*. So sound and cautious an investigator as Francis Galton had also in 1875 concluded that "acquired modifications are barely, if at all, *inherited*, in the correct sense of that word."

Darwin's belief in the inheritance of acquired characters was more or less hereditary in the family. His grandfather, Erasmus Darwin, anticipated Lamarck's views in his *Zoonomia*, which Darwin at one time "greatly admired." His father was "convinced" of the "inherited evil effects of alcohol," and to this extent at least he strongly impressed the belief in the inheritance [47] of acquired characters upon his children's minds. [16] Darwin must also have been imbued with Lamarckian ideas from other sources, although Dr. Grant's enthusiastic advocacy entirely failed to convert him to a belief in evolution. [17] "Nevertheless," he says, "it is probable that the hearing rather early in life such views maintained and praised may have favoured my upholding them under a different form in my *Origin of Species*"—a remark which refers to Lamarck's views on the general doctrine of evolution, but might also prove equally true if applied to Darwin's partial retention of the Lamarckian explanation of that evolution. Professor Huxley has pointed out that in [48] Darwin's earlier sketch of his theory of evolution (1844) he attached more weight to the inheritance of acquired habits than he does in his

Origin of Species published fifteen years later. [18] He appears to have acquired the belief in early life without first questioning and rigorously testing it as he would have done had it originated with himself. In later life it appeared to assist his theory of evolution in minor points, and in particular it appeared absolutely indispensable to him as the *only* explanation of the diminution of disused parts in cases where, as in domestic animals, economy of growth seemed to be practically powerless. He failed to adequately notice the effect of panmixia, or the withdrawal of selection, in causing or allowing degeneracy and dwindling under disuse; and he hardly attached sufficient importance to the fact that rudimentary organs and other supposed effects of use or [49] disuse are quite as marked features in neuter insects which cannot transmit the effects of use and disuse as they are in the higher animals.

REDUCED WINGS OF BIRDS OF OCEANIC ISLANDS.

Darwin himself has pointed out that the rudimentary wings of island beetles, at first thought to be due to disuse, are mainly brought about by natural selection—the best-winged beetles being most liable to be blown out to sea. But he says that in birds of the oceanic islands "not persecuted by any enemies, the reduction of their wings has probably been caused by disuse." This explanation may be as fallacious as it is acknowledged to have been in the case of the island beetles. According to Darwin's own views, natural selection *must* at least have played an important part in reducing the wings; for he [50] holds that "natural selection is continually trying to economize every part of the organization." He says: "If under changed conditions of life a structure, before useful, becomes less useful, its diminution will be favoured, for it will profit the individual not to have its nutriment wasted in building up an useless structure.... Thus, as I believe, natural selection will tend in the long run to reduce any part of the organization, as soon as it becomes, through changed habits, superfluous." [19] If, as Darwin powerfully urges (and he here ignores his usual explanation), ostriches' wings are insufficient for flight in consequence of the economy enforced by natural selection, [20] why may not the reduced wings of the dodo, or the penguin, or the apteryx, or of the Cursores generally, be wholly attributed to natural selection in favour of economy of material and adaptation of parts to changed conditions? [51] The great principle of economy is continually at work shaping organisms, as sculptors shape statues, by removing the superfluous parts; and a mere glance at the forms of animals in general will show that it is well-nigh as dominant and universal a principle as is that of the positive development of useful parts. Other causes, moreover besides actual economy, would favour shorter and more convenient wings on oceanic islands. In the first place, birds that were somewhat weak on the wing would be most likely to settle on an island and stay there. Shortened wings would then become advantageous because they would restrain fatal migratory tendencies or useless and perilous flights in which the birds that flew furthest would be most often carried away by storms and adverse winds. Reduced wings would keep the birds near the shelter and the food afforded by the island and its neighbourhood, and in some cases would become adapted to act [52] as fins or flappers for swimming under water in pursuit of fish.

The reduced size of the wings of these island birds is paralleled by the remarkable thinness, &c., of the shell of the "gigantic land-tortoise" of the Galapagos Islands. The changes seen in the carapace can hardly have been brought about by the inherited effects of special disuse. Why then should not the reduction of equally useless, more wasteful, and perhaps positively dangerous wings be also due to an economy which has become advantageous to bird and reptile alike through the absence of the mammalian rivals whose places they are evidently being modified to fill? The *complete* loss of the wings in neuter ants and termites can scarcely be due to the inherited effects of disuse; and as natural selection has abolished these wings in spite of the opposition of use-inheritance, it must clearly be fully competent to reduce wings without its aid. In considering the [53] rudimentary wings of the apteryx, or of the moa, emu, ostrich, &c., we must not forget the frequent or occasional occurrence of hard seasons, and times of drought and famine, when Nature eliminates redundant, wasteful, and ill-adapted organisms in so severe and wholesale a fashion. Where enemies are absent there would be unrestrained multiplication, and this would greatly increase the severity of the competition for food, and so hasten the elimination of disused and useless parts.

DROOPING EARS AND DETERIORATED INSTINCTS.

Mr. Galton has pointed out that existing races and existing organs are only kept at their present high pitch of organic excellence by the stringent and incessant action of natural or artificial selection; and the simple relaxation or withdrawal of such selective influences will almost necessarily [54] result in a certain amount of deterioration, independently even of the principle of economy. [21] I think that this cessation of a previous selective process will account for the drooping—but *not diminished*—ears of various domesticated animals (human preference and increased weight evidently aiding), and also for the inferior instincts seen in them and in artificially-fed caterpillars of the silk-moth, which now "often commit the strange mistake of devouring the base of the leaf on which they are feeding, and consequently fall down." Anyhow, I fail to see that anything is proved by this latter case, except that natural instinct may be perverted or aborted under unnatural conditions and a changed method of selection which abolishes the powerful corrective formerly supplied by natural selection.

[55]

WINGS AND LEGS OF DUCKS AND FOWLS.

The reduced wings and enlarged legs of domesticated ducks and fowls are attributed by Darwin and Spencer to the inheritance of the effects of use and disuse. But the inference by no means follows. Natural selection would usually favour these adaptive changes, and they would also have been aided by an artificial selection which is often unconscious or indirect. Birds with diminished power of flight would be less difficult to keep and manage, and in preserving and multiplying such birds man would be unconsciously bringing about structural changes which would easily be regarded as effects of use and disuse. "About eighteen centuries ago Columella and Varro speak of the necessity of keeping ducks in netted enclosures like other wild fowl, so that at this period there was danger of their flying away." [22] Is it not probable that the [56] best fliers would escape most frequently, or would pine most if kept confined? On the other hand, birds with lessened powers of flight would not be eliminated as under natural conditions, but would be favoured; and natural selection, together with artificial selection of the most flourishing birds, would thicken and strengthen the legs to meet increased demands upon them.

The diminution of the duck's wing is not great even in the birds that "never fly," and from this we must deduct the direct effect of disuse on the individual during its lifetime. As Weismann suggests, the *inherited* portion of the change could only be ascertained by comparing the bones, &c., of wild and tame ducks *similarly reared*. If individual disuse diminished the weight of the duck's wing-bones by 9 per cent. there would be nothing left to account for.

I suspect that investigation would reveal anomalies inconsistent with the theory of use-inheritance. [57] Thus according to Darwin's tables of comparative weights and measurements [23] the leg-bones of the Penguin duck have slightly diminished in length, although they have increased 39 per cent. in weight. Relatively to the weight of the skeleton, the leg-bones have shortened in the tame breeds of ducks by over 5 per cent. (and in two breeds by over 8 per cent.) although they have increased more than 28 per cent. in proportional weight. [24] How can increased use simultaneously shorten and thicken these bones? If the relative shortening is attributed to a heavier [58] skeleton, then the apparently reduced weight of the wing-bones is fully accounted for by the same circumstance, and disuse has had no inherited effect.

Another strange circumstance is that the wing-bones have diminished *in length only*. The shortening is about 6 per cent. more than in the shortened legs, and it amounts to 11 per cent. as compared with the weight of the skeleton. Such a shortening should represent a reduction of 29 per cent. in weight, whereas the actual reduction in the weight of the wing-bones relatively to the weight of the skeleton is only 9 per cent. even in the breeds that never fly. Independently of shortening, the disused wing-bones have actually thickened or increased in weight. In the Aylesbury duck the disproportion caused by these conflicting changes is so great that the wing-bones are 47 per cent. heavier than they should be if their weight had varied proportionally with their [59] length. [25] The reduction in weight on which Darwin relies seems to be entirely due to the shortening, and this shortening appears to be irrelevant to disuse, since the wings of the Call duck are similarly shortened in their proportions by 12 per cent., although this bird habitually flies to such an extent that Darwin partly attributes the greatly increased weight of its wing-bones to increased use under domestication.

We find that *all* the changes are in the direction of shorter and thicker bones—a tendency which must be largely dependent upon the suspension of the rigorous elimination which keeps the [60] bones of the wild duck *long and light*. The used leg-bones and the disused wing-bones have alike been shortened and thickened, though in different proportions. Natural or artificial selection might easily thicken legs without lengthening them, or shorten wings without eliminating strong heavy bones, but it can hardly be contended that use-inheritance has acted in such conflicting ways. The thickening of the wing-bones has actually more than kept pace with any increase of weight in the skeleton, in spite of the effect of individual disuse and of the alleged cumulative effect of ancestral disuse for hundreds of generations. The case of the duck deserves special attention as a crucial one, if only from the fact that in this instance, and in this instance only, has Darwin given the weights of the skeletons, thus furnishing the means for a closer examination of his details than is usually possible.

If we ignore such factors as selection, panmixia, [61] correlation, and the effects of use and disuse during lifetime, and still regard the case of the domestic duck as a valid proof of the inheritance of the effects of use and disuse, we must also accept it as an equally valid proof that the effects of use and disuse are *not* inherited. Nay, we may even have to admit that, in two points out of four, the *inherited* effect of use and disuse on successive generations is exactly opposite to the immediate effect on the individual.

Among fowls the wing-bones have lost much in weight but little or nothing in length—which is the reverse of what has occurred in ducks, although disuse is alleged to be the common cause in both cases. Some of the fowls which fly least have their wing-bones as long as ever. In the case of the Silk and Frizzled fowls—ancient breeds which "cannot fly at all"—and in that of the Cochins, which "can hardly fly up to a low perch," Darwin observes "how truly the proportions of an organ [62] may be inherited although not fully exercised during many generations." [26] In four out of twelve breeds the wing-bones had become slightly heavier relatively to the leg-bones. Do not these facts tend to show that the changes in fowls' wings are due to fluctuating variability and selective influences rather than to a general law whereby the effects of disuse are cumulatively inherited?

PIGEONS' WINGS.

Concerning pigeons' wings Darwin says: "As fancy pigeons are generally confined in aviaries of moderate size, and as even when not confined they do not search for their own food, they must during many generations have used their wings incomparably less than the wild rock-pigeon ... but when we turn to the wings we find what at [63] first appears a wholly different and unexpected result." [27] This unexpected increase in the spread of the wings from tip to tip is due to the feathers, which have lengthened in spite of disuse. Excluding the feathers, the wings were shorter in seventeen instances, and longer in eight. But as artificial selection has lengthened the wings in some instances, why may it not have shortened them in others? Wings with shortened bones would fold up more neatly than the long wings of the Carrier pigeon for instance, and so might unconsciously be favoured by fanciers. The selection of elegant birds with longer necks or bodies would cause a relative reduction in the wings—as with the Pouter, where the wings have been greatly lengthened but not so much as the body. [28] Slender bodies, too, and the lessened divergence of the furculum, [29] would [64] slightly diminish the spread of the wings, and so would affect the measurements taken. As the wing-bones, moreover, are to some extent correlated with the beak and the feet, the artificial selection of shortened beaks might tend to shorten the wing as well as the feet. Under these circumstances how can we be sure of the actual efficacy of use-inheritance? Surely selection is as fully competent to effect slight changes in the direction of use-inheritance as it undoubtedly is to effect great changes in direct opposition to that alleged factor of evolution.

SHORTENED BREAST-BONE IN PIGEONS.

The shortening of the sternum in pigeons is attributed to disuse of the flight muscles attached to it. The bone is only shortened by a third of an inch, but this represents a very remarkable reduction in proportional length, which Darwin estimates at [65] from one-seventh to one-eighth, or over 13 per cent. This marked reduction, too, quite unlike the slight reduction of the wing-bones to which the other ends of the muscles are attached, was universal in the eleven specimens measured by Darwin; and the bone, though acknowledged to have been modified by artificial selection in some breeds, is not so open to observation as wings or legs. Even, however, if this relative shortening of the sternum remained otherwise inexplicable, it might still be as irrelevant to use and disuse as is the fact that "many breeds" of fancy pigeons have lost a rib, having only seven where the ancestral rock-pigeon has eight. [30] But the excessive reduction in the sternum is far from being inexplicable. In the first place Darwin has somewhat over-estimated it. Instead of comparing the deficiency of length with the increased length which *should* have been acquired [66] (since the pigeons have increased in average size) he compares it with the length of the breast-bone in the rock-pigeon. [31] By this method if a pigeon had doubled in dimensions while its breast-bone remained unaltered, the reduction would be put down as 100 per cent., whereas obviously the true reduction would be one-half, or 50 per cent. of what the bone *should be*. Avoiding this error and a minor fallacy besides, a sound estimate reduces the supposed reduction of 13 or 14 per cent. to one of 11·7 per cent., which is still of course a considerable diminution.

Part of this reduction must be due to the direct effect of disuse during the lifetime of the individual. Another and perhaps very considerable part of the relative change must be attributed to the lengthening of the neck or body by [67] artificial selection, or to other modifications of shape and proportion effected directly or indirectly by the same cause. [32] The reduction is greatest in the Pouter (18½ per cent.) and in the Pied Scanderoon (17½ per cent.). In the former the body has been greatly elongated by artificial selection and three or four additional vertebræ have been acquired in the hinder part of the body. [33] In the latter a long neck increases the length of the bird, and so causes, or helps to cause, the relative shortening of the breast-bone. In the English Carrier — which experiences the effects of disuse, as it is too valuable to be flown — the relative reduction of 11 per cent. is apparently more than accounted [68] for by the "elongated neck." The Dragon also has a long neck. In the Pouter, although the breast-bone has been shortened by 18½ per cent. relatively to the length of the body, it has *lengthened* by 20 per cent. relatively to the *bulk* of the body. [34] Darwin forgot to ask whether allowance must not be made for a frequent, or perhaps general, elongation of the neck and the hinder part of the body, and the relative shortening or the throwing forward of the central portion containing the ribs (frequently one less in number) and the sternum. The whole body of the pigeon is so much under the control of artificial selection, that every precaution must be taken to guard against such possible sources of error. [35]

[69]

Under domestication there would be a suspension of the previous elimination of reduced breast-bones by natural selection (Weismann's panmixia), and a diminution of the parts concerned in flying might even be favoured, as lessened powers of *continuous* flight would prevent pigeons from straying too far, and would fit them for domestication or confinement. Such causes might reduce some of the less observed parts affected by flying, while still leaving the wing of full size for occasional flight, or to suit the requirements of the pigeon-fanciers. A change might thus be commenced like that seen in the rudimentary keel of the sternum in the owl-parrot of New Zealand, which has lost the power of flight although still retaining fairly-developed wings.

[70]

SHORTENED FEET IN PIGEONS.

Darwin thinks it highly probable that the short feet of most breeds of pigeons are due to lessened use, though he owns that the effects of correlation with the shortened beak are more plainly shown than the effects of disuse. [36] But why need the inherited effects of disuse be called in to explain an average reduction of some 5 per cent., when Darwin's measurements show that in the breeds where long beaks are favoured the principle of correlation between these parts has lengthened the foot by 13 per cent. in spite of disuse?

SHORTENED LEGS OF RABBITS.

In the case of the domestic rabbit Darwin notices that the bones of the legs have (relatively) become shorter by an inch and a half. But [71] as the leg-bones have *not* diminished in relative weight, [37] they must clearly have grown *thicker* or denser. If disuse has shortened them, as Darwin supposes, why has it also thickened them? The ears and the tail have been lengthened in spite of disuse. Why then may not the ungainly hind-legs have been shortened by human preference independently of the inherited effects of disuse? By relying on apparently favourable instances and neglecting the others it would be easy to arrive at all manner of unsound conclusions. We might thus become convinced that vessels tend to sail northwards, or that a pendulum oscillates more often in one direction than in the other. It must not be forgotten that it would be easy to cite an enormous number of cases which are in direct conflict with the supposed law of use-inheritance.

[72]

BLIND CAVE-ANIMALS.

Weak or defective eyesight is by no means rare as a spontaneous variation in animals, "the great French veterinary Huzard going so far as to say that a blind race [of horses] could soon be formed." Natural selection evolves blind races whenever eyes are useless or disadvantageous, as with parasites. This may apparently be done independently of the effects of disuse, for certain neuter ants have eyes which are reduced to a more or less rudimentary condition, and neuter termites are blind as well as wingless. In one species of ant (*Eciton vastator*) the sockets have disappeared as well as the eyes. In deep caves not only would natural selection cease to maintain good eyesight but it would persistently favour blindness — or the entire removal of the eye when greatly exposed, as in the cave-crab — and as Dr. Ray Lankester has [73] indicated, [38] there would have been a previous selection of animals which through spontaneous weakness, sensitiveness, or other affection of the eye found refuge and preservation in the cave, and a subsequent selection of the descendants whose fitness for relative darkness led them deeper into the cave or prevented them from straying back to the light with its various dangers and severer competition. Panmixia, however, as Weismann has shown, would probably be the most important factor in causing blindness.

INHERITED HABITS.

Darwin says: "A horse is trained to certain paces, and the colt inherits similar consensual movements." [39] But selection of the constitutional [74] tendency to these paces, and imitation of the mother by the colt, may have been the real causes. The evidence, to be satisfactory, should show that such influences were excluded. Men acquire proficiency in swimming, waltzing, walking, smoking, languages, handicrafts, religious beliefs, &c., but the children only appear to inherit the innate abilities or constitutional proclivities of their parents. Even the songs of birds, including their call-notes, are no more inherited than is language by man (*Descent of Man*, p. 86). They are learned from the parent. Nestlings which acquire the song of a distinct species, "teach and transmit their new song to their offspring." If use-inheritance has not fixed the song of birds, why should we suppose that in a single generation it has transmitted a newly-taught method of walking or trotting?

It is alleged that dogs inherit the intelligence acquired by association with man, [75] and that retrievers inherit the effects of their training. [40] But selection and imitation are so potent that the additional hypothesis of use-inheritance seems perfectly superfluous. Where intelligence is not highly valued and carefully promoted by selection, the intelligence derivable from association with man does *not* appear to be inherited. Lap-dogs, for instance, are often remarkably stupid.

Darwin also instances the inheritance of dexterity in seal-catching as a case of use-inheritance. [41] But this is amply explained by the ordinary law of heredity. All that is needed is that the son shall inherit the suitable faculties which the father inherited before him.

[76]

TAMENESS OF RABBITS.

Darwin holds that in some cases selection alone has modified the instincts and dispositions of domesticated animals, but that in most cases selection and the inheritance of acquired habits have concurred in effecting the change. "On the other hand," he says, "habit alone in some cases has sufficed; hardly any animal is more difficult to tame than the young of the wild rabbit; scarcely any animal is tamer than the young of the tame rabbit; but I can hardly suppose that domestic rabbits have often been selected for tameness alone; so that we must attribute at least the greater part of the inherited change from extreme wildness to extreme tameness to habit and long-continued close confinement." [42]

But there are strong, and to me irresistible, arguments to the contrary. I think that the following [77] considerations will show that the greater part, if not the whole, of the change must be attributed to selection rather than to the direct inheritance of acquired habit.

(1) For a period which may cover thousands of generations, there has been an entire cessation of the natural selection which maintains the wildness (or excessive fear, caution, activity, &c.) so indispensably essential for preserving defenceless wild rabbits of all ages from the many enemies that prey upon them.

(2) During this same extensive period of time man has usually killed off the wildest and bred from the tamest and most manageable. To some extent he has done this consciously. "It is very conducive to successful breeding to keep only such as are quiet and tractable," says an authority on rabbits, [43] and he enjoins the selection of the [78] handsomest and *best-tempered* does to serve as breeders. To a still greater extent man has favoured tameness unconsciously and indirectly. He has systematically selected the largest and most prolific animals, and has thus doubled the size and the fertility of the domestic rabbit. In consciously selecting the largest and most flourishing individuals and the best and most prolific mothers, he *must* have unconsciously selected those rabbits whose relative *tameness* or placidity of disposition rendered it possible for them to flourish and to produce and rear large and thriving families, instead of fretting and pining as the wilder captives would do. When we consider how exceedingly delicate and easily disturbed yet all-

important a function is that of maternity in the continually breeding rabbit, we see that the tamest and the least terrified would be the most successful mothers, and so would continually be selected, although man cared nothing for the tameness [79] in itself. The tamest mothers would also be less liable to neglect or devour their offspring, as rabbits commonly do when their young are handled too soon, or even when merely frightened by mice, &c., or disturbed by changed surroundings.

(3) We must remember the extraordinary fecundity of the rabbit and the excessive amount of elimination that consequently takes place either naturally or artificially. Where nature preserved only the wildest, man has preserved the tamest. If there is any truth in the Darwinian theory, this thorough and long-continued reversal of the selective process *must* have had a powerful effect. Why should it not be amply sufficient to account for the tameness and mental degeneracy of the rabbit without the aid of a factor which can readily be shown to be far weaker in its normal action than either natural or artificial selection? Why may not the tameness of the rabbit be transferred [80] to the group of cases in which Darwin holds that "habit has done nothing," and selection has done all?

(4) If use-inheritance has tamed the rabbit, why are the bucks still so mischievous and unruly? Why is the Angora breed the only one in which the males show no desire to destroy the young? Why, too, should use-inheritance be so much more powerful in the rabbit than with other animals which are far more easily tamed in the first instance? Wild young rabbits when domesticated "remain unconquerably wild," and, although they may be kept alive, they pine and "rarely come to any good." Yet the animal which *acquires* least tameness—or apparently, indeed, none at all—inherits most! It appears, in fact, to inherit that which it cannot acquire—a circumstance which indicates the selection of spontaneous variations rather than the inheritance of changed habits. Such variations occasionally occur in animals in a [81] marked degree. Of a litter of wolf-cubs, all brought up in the same way, "one became tame and gentle like a dog, while the others preserved their natural savagery." Is it not probable that permanent domestication was rendered possible by the inevitable selection of spontaneous variations in this direction? The *excessive* tameness, too, of the young rabbit, while easily expli-

cable as a result of unconscious selection, is not easily explained as a result of acquired habit. No particular care is taken to tame or teach or domesticate rabbits. They are bred for food, or for profit or appearance, and they are left to themselves most of their time. As Sir J. Sebright notices with some surprise, the domestic rabbit "is not often visited, and seldom handled, and yet it is always tame."

[82]

MODIFICATIONS OBVIOUSLY ATTRIBUTABLE TO SELECTION.

Innumerable modifications in accordance with altered use or disuse, such as the enlarged udders of cows and goats, and the diminished lungs and livers in highly bred animals that take little exercise, can be readily and fully explained as depending on selection. As the fittest for the natural or artificial requirements will be favoured, natural or artificial selection may easily enlarge organs that are increasingly used and economize in those that are less needed. I therefore see no necessity whatever for calling in the aid of use-inheritance as Darwin does, to account for enlarged udders, or diminished lungs, or the thick arms and thin legs of canoe Indians, or the enlarged chests of mountaineers, or the diminished eyes of moles, or the lost feet of certain beetles, or the reduced wings of logger-headed ducks, or the prehensile tails of [83] monkeys, or the displaced eyes of soles, or the altered number of teeth in plaice, or the increased fertility of domesticated animals, or the shortened legs and snouts of pigs, or the shortened intestines of tame rabbits, or the lengthened intestines of domestic cats, &c. [44] Changed habits and the requisite change of structure will usually be favoured by natural selection; for habit, as Darwin says, "almost implies that some benefit great or small is thus derived."

SIMILAR EFFECTS OF NATURAL SELECTION AND USE-INHERITANCE.

Here we perceive a difficulty which will equally trouble those who affirm use-inheritance and those [84] who deny. Broadly speaking, the adaptive effects ascribed to use-inheritance coincide with the effects of natural selection. The individual adaptability (as shown in the thickening of skin, fur, muscle, &c., under the stimulus of friction, cold, use, &c.) is identical in kind and direction with the racial adaptability under natural selection. Consequently the alleged inheritance of the advantageous effects of use and disuse cannot readily be distinguished from the similarly beneficial effects of natural selection. The indisputable fact that natural selection imitates or simulates the beneficial effects ascribed to use-inheritance may be the chief source and explanation of a belief which may prove to be thoroughly fallacious. A similar simulation of course occurs under domestication, where natural selection is partly replaced by artificial selection of the best adapted and therefore most flourishing animals, while in disused parts panmixia or the comparative cessation of selection will aid or [85] replace "economy of growth" in causing diminution. [45]

INFERIORITY OF SENSES IN EUROPEANS.

"The inferiority of Europeans, in comparison with savages, in eyesight and in the other senses," is attributed to "the accumulated and transmitted effect of lessened use during many generations." [46] But why may we not attribute it to the slackened and diverted action of the natural selection which keeps the senses so keen in some savage races?

SHORT-SIGHT IN WATCHMAKERS AND ENGRAVERS.

Darwin notices that watchmakers and engravers are liable to be short-sighted, and that short-sight [86] and long-sight certainly tend to be inherited. [47] But we must be careful not to beg the question at issue by assuming that the frequent heredity of short sight necessarily covers the heredity of artificially-produced short-sight. Elsewhere, however, Darwin states more decisively that "there is ground for believing that it may often originate in causes acting on the individual affected, and may thence-forward become transmissible." [48] This impression may arise (1) from the facts of ordinary heredity—the ancestral liability being excited in father and son by similar artificial habits, such as reading, and viewing objects closely as among watchmakers and engravers—or by constitutional deterioration from indoor life, &c., acting upon a constitutional liability of the eye to the "something like inflammation of the coats, under which they yield" and so [87] cause shortness of sight by altering the spherical shape of the eye-ball. (2) Panmixia, or the suspension of natural selection, together with altered habits, will account for an increase of short-sight among the population generally. (3) Long-sighted people could not work at watchmaking and engraving so comfortably and advantageously as at other occupations, and hence would be less likely to take to such callings.

LARGER HANDS OF LABOURERS' INFANTS. [49]

These are best explained as the result of natural selection and of the diminution of the hand by sexual selection in the gentry. If the larger hands of labourers' infants are really due to the inherited effects of ancestral use, why does the development occur so early in life, instead of only at a corresponding period, as is the rule? During the first [88] few years of its life, at least, the labourer's infant does no more work than the gentleman's child. Why are not the effects of this disuse inherited by the labourer's infant? If the enlargement of the infant's hand illustrates the transference of a character gained later in life, it is evident that the transference must take place in spite of the inherited effects of disuse.

THICKENED SOLE IN INFANTS.

Darwin also attributes the thickened sole in infants, "long before birth," to "the inherited effects of pressure during a long series of generations." [50] But disuse should make the infant's sole *thin*, and it is this thinness that should be inherited. If we suppose the inheritance of the thickened soles of later life to be transferred to an earlier period, we have the anomaly of the inherited effects of disuse [89] at that earlier period being overpowered by the untimely inheritance of the effects of use at another. On the other hand, it is clear that natural selection would favour thickened soles for walking on, and might also promote an early development which would ensure their being ready in good time for actual use; for variations in the direction of delay would be cut off, while variations in the other direction would be preserved. Anyhow, the mere transference of a character to an earlier period is no proof of use-inheritance. The real question is whether the thickened sole was gained by natural selection or by the inherited effects of pressure, and the mere transference or hastened appearance of the thickening does not in any degree solve this question. It merely excludes the effect of disuse during lifetime, and thus presents a fallacious appearance of being decisive. The thickened sole of the unborn infant, however, like the lanugo or hairy covering, is probably a result of the direct [90] inheritance of ancestral stages of evolution, of which the embryo presents a condensed epitome. While the relative thinness of the infant's sole might be pointed to as the effect of *disuse* during a long series of generations, its thickness is rather an illustration of atavism still resisting the effects of long-continued disuse. There is nothing to show that the inheritable portion of the full original thickness was not gained by natural selection rather than by the directly inherited effect of use; and the latter, being cumulative and indiscriminative in its action, would apparently have made the sole very much thicker and harder than it is. If natural selection were not supreme in such cases, how could we account for the effects of pressure resulting in hard hoofs in some cases and only soft pads in others?

[91]

A SOURCE OF MENTAL CONFUSION.

Of course in a certain sense this thickening of the sole has resulted from use. In one sense or other, most — or perhaps all — of the results of natural selection are inherited effects of use or disuse. Natural selection preserves that which is of use and which is used, while it eliminates that which is useless and is not used. The most confident assertions of the effects of use and disuse in modifying the heritable type, appear to rest on this indefeasible basis. Darwin's statements concerning the effects of use and disuse in evolution can frequently be read in two senses. They often command assent as undeniable truisms as they stand, but are of course written in another and more debatable sense. Thus in the case of the shortened wings and thickened legs of the domestic duck, I believe equally with Darwin and Spencer that "no one will dispute that they have resulted [92] from the lessened use of the wings and the increased use of the legs." "Use" is at bottom the determining circumstance in evolution generally. The trunk of the elephant, the fin of the fish, the wing of the bird, the cunning hand of man and his complicated brain—and, in short, all organs and faculties whatsoever—can only have been moulded and developed by use—by usefulness and by using—but not necessarily by use-inheritance, not necessarily by directly inherited effects of use or disuse of parts in the individual. So, too, reduced or rudimentary organs are due to disuse, but it by no means follows that the diminution is caused by any direct tendency to the inheritance of the effects of disuse in the individual. The effects of natural selection are commonly expressible as effects of use and disuse, just as adaptation in nature is expressible in the language of teleology. But use-inheritance is no more proven by one of these necessary coincidences than special design is by the other. [93] The inevitable simulation of use-inheritance may be entirely deceptive.

Darwin thinks that "there can be no doubt that use in our domestic animals has strengthened and enlarged certain parts, and disuse diminished them; and that such modifications are inherited." Undoubtedly "such" or *similar* modifications have often been inherited, but how can Darwin possibly tell that they are not due to the simulation of use-inheritance by natural or artificial selection acting upon general variability? Of the inevitability of selection and of its generally adaptive tendencies "there can be no doubt," and panmixia would tend to reduce disused parts; so that there *must always* remain grave doubts of the alleged inheritance of the similar effects of use and disuse, unless we can accomplish the extremely difficult feat of excluding both natural and artificial selection as causes of enlargement, and panmixia and selection as causes of dwindling.

[94]

WEAKNESS OF USE-INHERITANCE.

Use-inheritance is normally so weak that it appears to be quite helpless when opposed to any other factor of evolution. Natural selection evolves and maintains the instincts of ants and termites in spite of use-inheritance to a more wonderful degree than it evolves the instincts of almost any other animal with the fullest help of use-inheritance. It develops seldom-used horns or natural armour just as readily as constantly-used hoofs or teeth. Sexual selection evolves elaborate structures like the peacock's tail in spite of disuse and natural selection combined. Artificial selection appears to enlarge or diminish used parts or disused parts with equal facility. The assistance of use-inheritance seems to be as unnecessary as its opposition is ineffective.

The alleged inheritance of the effects of use and disuse in our domestic animals must be very [95] slow and slight. [51] Darwin tells us that "there is no good evidence that this ever follows in the course of a single generation." "Several generations [96] must be subjected to changed habits for any appreciable result." [52] What does this mean? One of two things. Either the tendency is very weak, or it is non-existent. If it is so weak that we cannot detect its alleged effects till several generations have elapsed, during which time the more powerful agency of selection has been at work, how are we to distinguish the effects of the minor factor from that of the major? Are we to conclude that use-inheritance *plus* selection will modify races, just as Voltaire firmly held that incantations, together with sufficient arsenic, would destroy flocks of sheep? Is it not a significant fact that the alleged instances of use-inheritance so often prove to be self-conflicting in their details?

For satisfactory proof of the prevalence of a law of use-inheritance we require normal instances [97] where selection is clearly inadequate to produce the change, or where it is scarcely allowed time or opportunity to act, as in the immediate offspring of the modified individual. Of the first kind of cases there seems to be a plentiful lack. Of the latter kind, according to Darwin, there appears to be none — a circumstance which contrasts strangely and suspiciously with the many decisive cases in which variation from unknown causes has been inherited most strikingly in the immediate offspring. It must be expected, indeed, that among these innumerable cases some will accidentally mimic the alleged effects of use-inheritance.

If Darwin had felt certain that the effects of habit or use tended in any marked degree to be conveyed directly and cumulatively to succeeding generations, he could hardly have given us such cautious, half-hearted encouragement of good habits as the following: — "It [98] is not improbable that after long practice virtuous tendencies may be inherited." "Habits, moreover followed during many generations probably tend to be inherited." [53] This is probable, independently of use-inheritance. The "many generations" specified or implied, will allow time for the play of selective as well as of cumulatively-educative influences. There must apparently be a constitutional or inheritable predisposition or fitness for the habits spoken of, which otherwise would scarcely be continued for many generations, except by the favourably-varying branches of a family: which again is selection rather than use-inheritance.

Where is the necessity for even the remains of the Lamarckian doctrine of inherited habit? Seeing how powerful the general principle of selection has shown itself in cases where use-inheritance could have given no aid or must [99] even have offered its most strenuous opposition, why should it not equally be able to develop used organs or repress disused organs or faculties without the assistance of a relatively weak ally? Selection evolved the remarkable protective coverings of the armadillo, turtle, crocodile, porcupine, hedgehog, &c.; it formed alike the rose and its thorn, the nut and its shell; it developed the peacock's tail and the deer's antlers, the protective mimicry of various insects and butterflies, and the wonderful instincts of the white ants; it gave the serpent its deadly poison and the violet its grateful odour; it painted the gorgeous plumage of the Impeyan pheasant and the beautiful colours and decorations of countless birds and insects and flowers. These, and a thousand other achievements, it has evidently accomplished without the help of use-inheritance. Why should it be thought incapable of reducing a pigeon's wing or enlarging [100] a duck's leg? Why should it be credited with the help of an officious ally in effecting comparatively slight changes, when great and striking modifications are effected without any such aid?

FOOTNOTES:

[15] Weismann's *Essays on Heredity*, &c. Clarendon Press, 1889.

[16] *Life and Letters*, i. p. 16. Darwin's reverence for his father "was boundless and most touching. He would have wished to judge everything else in the world dispassionately, but anything his father had said was received with almost implicit faith; ... he hoped none

of his sons would ever believe anything because he said it, unless they were themselves convinced of its truth—a feeling in striking contrast with his own manner of faith" (*Life and Letters*, i. pp. 10, 11).

[17] *Ibid.*, i. p. 38.

[18] *Life and Letters*, ii. p. 14.

[19] *Origin of Species*, pp. 117, 118.

[20] *Ibid.*, p. 180.

[21] *Contemporary Review*, December, 1875, pp. 89, 93.

[22] *Variation of Animals and Plants under Domestication*, i. 292.

[23] *Variation of Animals and Plants under Domestication*, i. 299-301.

[24] To keep pace with this lateral increase in weight, the leg-bones should have lengthened considerably so that their total deficiency in proportional length is 17 per cent.,—a changed proportion which being *linear* is more excessive than the increase of weight by 28 per cent. So marked is the effect of the combined thickening and shortening that in the Aylesbury breed—which is the most typically representative one—the leg-bones have become 70 per cent. heavier than they should be if their thickness had continued to be proportional to their length.

[25] This excessive thickening under disuse appears to be due partly to a positive lateral enlargement or increase of proportional weight of about 7½ per cent., and partly to a shortening of about 15 per cent. Carefully calculated, the reduction of the weight of the wing-bones in this breed is only 8·3 per cent. relatively to the whole skeleton, or only 5 per cent. relatively to the skeleton *minus* legs and wings. The latter method is the more correct, since the excessive weight of the leg-bones increases the weight of the skeleton more than the diminished weight of the wing-bones reduces it.

[26] *Variation of Animals and Plants under Domestication*, i. 284.

[27] *Variation of Animals and Plants under Domestication*, i. 184, 185.

[28] *Ibid.*, i. 144, 145.

[29] *Ibid.*, i. 185.

[30] *Variation of Animals and Plants under Domestication*, i. 175.

[31] *Variation of Animals and Plants under Domestication*, i. 184. I suspect that Darwin was in poor health when he wrote this page. He nods at least four times in it. Twice he speaks of "twelve" breeds where he obviously should have said eleven.

[32] If a prominent breast is admired and selected by fanciers, the sternum might shorten in assuming a more forward and vertical position. If the shortening of the sternum is entirely due to disuse, it seems strange that Darwin has not noticed any similar shortening in the sternum of the duck. But selection has not tended to make the duck elegant, or "pigeon-breasted"; it has enlarged the abdominal sack instead, besides allowing the addition of an extra rib in various cases.

[33] *Variation of Animals and Plants under Domestication*, 144, 175.

[34] *Variation of Animals and Plants under Domestication*, i. 179.

[35] In the six largest breeds the shortening of the sternum is nearly twice as great as in the three smaller breeds which remain nearest the rock-pigeon in size. We can hardly suppose that use-inheritance especially affects the eight breeds that have varied most in size. If we exclude these, there is only a total shortening of 7 per cent. to be accounted for.

[36] *Variation of Animals and Plants under Domestication*, i. 183, 186.

[37] *Variation of Animals and Plants under Domestication*, i. 130, 135; ii. 288.

[38] *Encyclopædia Britannica*, article "Zoology."

[39] *Variation of Animals and Plants under Domestication*, ii. 367.

[40] *Variation of Animals and Plants under Domestication*, ii. 367. Why then does the cheetah inherit ancestral habits so inadequately that it is useless for the chase unless it has first learned to hunt for itself before being captured? (ii. 133).

[41] *Descent of Man*, p. 33.

[42] *Origin of Species*, pp. 210, 211.

[43] E. S. Delamer on *Pigeons and Rabbits*, pp. 132, 103. For other points referred to, see pages 133, 102, 100, 95, 131.

[44] *Origin of Species*, pp. 188, 110; *Descent of Man*, pp. 32-35; *Variation of Animals and Plants under Domestication*, ii. 289, 293. Use or disuse during lifetime of course co-operates, and in some cases, as in that of the canoe Indians, may be the principal or even perhaps the *sole* cause of the change.

[45] For the importance of panmixia as invalidating Darwin's strongest evidence for use-inheritance—namely, that drawn from the effects of disuse in highly-fed domestic animals where there is supposed to be no economy of growth—see Professor Romanes on Panmixia, *Nature*, April 3, 1890.

[46] *Descent of Man*, p. 33.

[47] *Descent of Man*, p. 33.

[48] *Variation of Animals and Plants under Domestication*, i., 453.

[49] *Descent of Man*, p. 33.

[50] *Descent of Man*, p. 33.

[51] Wallace shows that the changes in our domestic animals, if spread over the thousands of years since the animals were first tamed, must be extremely insignificant in each generation, and he concludes that such infinitesimal effects of use and disuse would be swallowed up by the far greater effects of variation and selection (*Darwinism*, p. 436). Professor Romanes has replied to him in the *Contemporary Review* (August 1889), showing that this is no disproof of the existence of the minor factor, inasmuch as slight changes in each generation need not necessarily be matters of life and death to the individual, although their cumulative development by use-inheritance might eventually become of much service. But selection would favour spontaneous variations of a similarly serviceable character. The slightest tendency to eliminate the extreme variations in either direction would proportionally modify the average in a breed. Use-inheritance appears to be so relatively weak a factor that probably neither proof nor disproof of its existence can ever be given, owing to the practical impossibility of disentangling its effects (if any) from the effects of admittedly far more powerful factors which often act in unsuspected ways. Thus wild ducklings, which can easily be reared by themselves, invariably "die off" if reared with tame ones (*Variation*, &c., i. 292, ii. 219). They cannot get their fair

share in the competition for food, and are completely eliminated. Professor Romanes fully acknowledges that there is the "gravest possible doubt" as to the transmission of the effects of disuse (Letter on Panmixia, *Nature*, March 13, 1890).

[52] *Variation of Animals and Plants under Domestication*, ii. 287-289.

[53] *Descent of Man*, pp. 612, 131.

[101]

INHERITED INJURIES.

INHERITED MUTILATIONS.

The almost universal *non-inheritance* of mutilations seems to me a far more valid argument *against* a general law of modification-inheritance than the few doubtful or abnormal cases of such inheritance can furnish in its favour. No inherited effect has been produced by the docking of horses' tails for many generations, or by a well-known mutilation which has been practised by the Hebrew race from time immemorial. As lost or mutilated parts are reproduced in offspring independently of the existence of those parts in the parent, there is the less reason [102] to suppose that the particular condition of parental parts transmits itself, or tends to transmit itself, to the offspring. So unsatisfactory is the argument derivable from inherited mutilations that Mr. Spencer does not mention them at all, and Darwin has to attribute them to a special cause which is independent of any general theory of use-inheritance. [54]

Darwin's most striking case—and to my mind the only case of any importance—is that of Brown-Séquard's epileptic guinea-pigs, which inherited the mutilated condition of parents who had gnawed off their own gangrenous toes when anæsthetic through the sciatic nerve having been divided. [103] [55] Darwin also mentions a cow that lost a horn by accident, followed by suppuration, and subsequently produced three calves which had on the same side of the head, instead of a horn, a bony lump attached merely to the skin. Such cases may seem to prove that mutilation *associated with morbid action* is occasionally inherited or repeated with a promptitude and thoroughness that contrast most strikingly with the imperceptible nature of the immediate inheritance of the effects of use and disuse; but they by no means prove that mutilation in general is inheritable, and they are absolutely no proof whatever of a *normal* and non-pathological tendency to the inheritance of acquired characters. Those who accept Darwin's special explanation [104] of the supposed inheritance of mutilations, ought to notice that his explanation applies equally well under a theory which is strongly adverse to use-inheritance—namely, Galton's idea of the sterilization and complete "using up" of otherwise reproductive matter in the growth and maintenance of the personal structure.

Darwin's explanation of inherited mutilations—which, as he notes, occur "especially or perhaps exclusively" when the injury has been followed by disease [56]—is that all the representative gemmules which would develop or repair or reproduce the injured part are attracted to the diseased surface during the reparative process and are there destroyed by the morbid action. [105] [57] Hence they cannot reproduce the part in offspring. This explanation by no means implies that mutilation would *usually* affect the offspring. On the contrary, in all ordinary cases of mutilation the purely atavistic elements or gemmules would be set free from any modifying influence of the non-existent or mutilated part. The gemmules—as in Galton's theory of heredity and with neuter insects—might be perfectly independent of pangenesis and the normal inheritance of acquired characters. Such self-multiplying gemmules without pangenesis would enable us to understand both the excessive weakness or non-existence of normal use-inheritance, and the excessive strength and abruptness of the effect of their partial destruction under special pathological conditions.

The series of epileptic phenomena that can be excited by tickling a certain part of the cheek and neck of the adult guinea-pig during the growth and rejoining of the ends of the severed nerve, [106] are said to be repeated with striking accuracy of detail in the young who inherit mutilated toes; but as epilepsy is often due to some *one* exciting cause or morbid condition, the single transmission of a highly morbid condition of the system might easily reproduce the whole chain of consequences and might also have caused the loss of toes.

The particulars of the guinea-pig cases are very inadequately recorded, [58] but the results are so anomalous [59] that Brown-Séquard's own conclusion [107] is that the epilepsy and the inherited injuries are *not* directly transmitted, but that "what is transmitted is the morbid state of the nervous system." He thinks that the missing toes may "possibly" be exceptions to this conclusion, "but the other facts only imply the transmission of a morbid state of the sympathetic or sciatic nerve or of a part of the medulla oblongata." Until we can tell what is transmitted, we are not in a position to determine whether there is any true inheritance or only an exaggerated simulation of it under peculiar circumstances. When the actual observers believe that the mutilations and epilepsy are not the cause of their own repetition, [108] and when these observers guard themselves by such phrases as, "if any conclusion can at present be drawn from those facts," we who have only incomplete reports to guide us may well be excused if we preserve an even more pronounced attitude of caution and reserve. [60] The morbid state of the system may be wholly due to general injury of the germs rather than to specific inheritance.

Weismann suggests that the morbid condition of the nervous system may be due to some infection such as might arise from microbes, which find a home in the mutilated and disordered nervous system in the parent, and subsequently transmit themselves to the offspring through the reproductive elements, as the infections of various diseases appear to do—the muscardine silkworm [109] disease in particular being known to be conveyed to offspring in this manner.

But whether we can discover the true explanation or not, inherited mutilations can hardly be accounted for as the result of a general tendency to inherit acquired modifications. How could a factor which seems to be totally inoperative in cases of ordinary mutilation, and only infinitesimally operative in transmitting the normal effects of use and disuse, suddenly become so powerful as to completely overthrow atavism, and its own tendency to transmit the non-mutilated type of one of the parents and of the non-mutilated type presented by the injured parent in earlier life? Does not so striking and abrupt an intensification of its usually insignificant power demand an explanation widely different from that which might account for the extremely slow and slight inheritance of the normal effects of use and disuse? Surely it would be better to [110] suspend one's judgment as to the true explanation of highly exceptional and purely pathological cases rather than resort to an hypothesis that creates more difficulties than it solves.

THE MOTMOT'S TAIL.

The narrowing of the long central tail feathers of the motmot is attributed to the inherited effects of habitual mutilation (*Descent of Man*, pp. 384, 603). But in the specimens at South Kensington [61] the narrowness extends upwards much beyond the habitually denuded part, and the broadened end is the broadest part of the whole feather. If the inherited effect of an inch or two of denudation extends from three to six inches upwards, why has it not also extended two inches downwards so as to narrow the broadened end? [111] The narrowness seems to be a mainly relative or negative effect produced by the broadening out of a long tapering feather at its end under the influence of sexual selection. Several other birds have similarly narrowed or spoon-shaped feathers and do not bite them. Is it not more feasible to suppose that this attractive peculiarity first suggested its artificial intensification, than to suppose that the bird began nibbling without any definite cause? Sexual selection would then encourage the habit. Anyhow, it is as impossible to show that the mutilation preceded the narrowing as it is to show that tonsure preceded baldness.

OTHER INHERITED INJURIES MENTIONED BY DARWIN.

Darwin quotes some cases from Dr. Prosper Lucas's "long" but weak and unsatisfactory "list [112] of inherited injuries." [62] But Lucas was somewhat credulous. One of his cases is that many girls were born in London without mammæ through the injurious effect of certain corsets on the mothers. He also gives a long account of a Jew who could read through the thick covers of a book, and whose son inherited this "hyperæsthesia" of the sense of sight in a still more remarkable degree (i. 113-119). Evidently Lucas's cases cannot be accepted without some amount of reserve.

The cases of the three calves which inherited the one-horned condition of the cow, the two sons who inherited a father's crooked finger, and the two sons who were microphthalmic on the same side as their father had lost an eye, may be due to mere coincidence; or an inherited constitutional tendency or liability might lead to somewhat similar [113] results in parent and offspring [63]—just as the tendency to certain fatal diseases or to suicide may produce similar results in father and son, although the artificially-produced hanging or apoplexy obviously cannot be directly transmitted. That more than one of the offspring was affected does not render the chances against coincidence "almost infinitely great," as Darwin mistakenly supposes. It "frequently occurs" that a man's sons or daughters may *all* exhibit either a latent or a newly-developed congenital peculiarity previously unknown; [64] and the coincidence may merely be that one of the parents accidentally suffered a similar kind of injury—a kind of coincidence which must of course occasionally occur, and which may have been partly caused by a latent tendency. The chances [114] against coincidence are indeed great, but the cases appear to be correspondingly rare.

Darwin acknowledges that many supposed instances of inherited mutilation may be due to coincidence; and there is apparently no more reason for attributing inherited scars, &c., to any special form of heredity than to the effect of the mother's imagination on the unborn babe—a popular but fallacious belief in corroboration of which far more alleged instances could be collected than of the inheritance of injuries.

As an instance of the coincidences that occur, I may mention that a friend of mine has a daughter who was born with a small hole in one ear, just as if it were already pierced for the earring which she has since worn in it. I suppose, however, that no one will venture to claim this as an instance of the inheritance of a mutilation practised by female ancestors, especially as such holes are not altogether unknown or [115] inexplicable, though very rarely occurring low down in the lobe of the ear. [65]

Many cases are known of the inheritance of mutilations or malformations arising congenitally from some abrupt variation in the reproductive elements. In such cases as the one-eared rabbits, the two-legged pigs, the three-legged dogs, the one-horned stags, hornless bulls, earless rabbits, lop-eared rabbits, tailless dogs, &c., if the father or the mother or the embryo had suffered from some accident or disease which might plausibly have been assigned as the cause of the original malformation, these transmitted defects would readily be cited as instances of the inheritance of an accidentally-produced modification.

The inheritance of exostoses on horses' legs may be the inheritance of a constitutional tendency [116] rather than of the effect of the parents' hard travelling. Horses congenitally liable to such formations would transmit the liability, [66] and this might readily be mistaken for inheritance of the results of the liability. An apparent increase in this liability might arise from greater attention being now paid to it, or from increased use of harder roads; or a real increase might be due to panmixia and some obscure forms of correlation.

QUASI-INHERITANCE.

Of course artificially-caused ill-health or weakness in parents will tend in a general way to injure the offspring. But deterioration thus caused is only a form of quasi-inheritance, as I should prefer to call it. Semi-starvation in a new-born babe is *not* truly inherited from its half-starved mother, but is the direct result of insufficient [117] nourishment. The general welfare of germs—as of parasites—is necessarily bound up with that of the organism which feeds and shelters them, but this is not heredity, and is quite irrelevant to the question whether particular modifications are transmitted or not.

Another form of quasi-inheritance is seen in the communication of certain infections to offspring. Not being transmitted by the action of the organism so much as in defiance of it, such diseases are not truly hereditary, though for convenience' sake they are usually so described.

A perversion or prevention of true inheritance is also seen in the action of alcohol, or excessive overwork, or any other cause which by originating morbid conditions in individuals may also injure the reproductive elements.

These forms of quasi-inheritance are, of course, highly important so far as the improvement of the race is concerned. So, too, is the fact that [118] improved or deteriorated habits and thoughts are transmitted by personal teaching and influence and are cumulative in their effect. But all this must not be confounded with the inheritance of acquired characters. Cases of quasi-inheritance may perhaps be most readily distinguished from cases of true inheritance by the time test. When a modification acquired in adult life is promptly communicated to the child in early life or from birth, it may rightly be suspected that the inheritance, like that of money or title, is not truly congenital, but is extraneous or even anti-congenital in its nature. Judged by such a standard, the inherited injuries in Brown-Séquard's guinea-pigs are only exceptional cases of quasi-inheritance, and are not necessarily indicative of any general rule affecting true inheritance.

FOOTNOTES:

[54] A very able anatomist of my acquaintance denies the inheritance of mutilations and injuries, although he strongly believes in the inheritance of the effects of use and disuse.

[55] *Variation of Animals and Plants under Domestication*, i. 467-469. Lost toes were only seen by Dr. Dupuy in three young out of two hundred. Obersteiner found that most of the offspring of his epileptic guinea-pigs were injuriously affected, being weakly, small, paralysed in one or more limbs, and so forth. Only two were epileptic, and both were weakly and died early (Weismann's *Essays*, p. 311). A morbid condition of the spinal cord might affect the hind limbs especially (as in paraplegia) and might occasionally cause loss of toes in the embryo by preventing development or by ulceration. Brown-Séquard does not say that the defective feet were on the same side as in the parents (*Lancet*, Jan., 1875, pp. 7, 8).

[56] *Variation of Animals and Plants under Domestication*, ii. 57.

[57] *Ibid.*, ii. 392. Perhaps it might be better to suppose that the *best* gemmules were sacrificed in repairing the injured *nerve*, and hence only inferior substitutes were left to take their place, and could only imperfectly reproduce the injured part of the nervous system in offspring.

[58] Hence perhaps Mr. Spencer's error in representing the epileptic liability as permanent and as coming on *after* healing (*Factors of Organic Evolution*, p. 27).

[59] It is not claimed that the imperfect foot was on the same side of the body as in the parent, and where parents had lost *all* the toes

of a foot, or the whole foot, the few offspring affected usually had lost only two toes out of the three, or only a part of one or two or three toes. Sometimes the offspring had toes missing on *both* hind feet, although the parent was only affected in *one*. *One* diseased ear and eye in the parent was "generally" or "always" succeeded by *two* equally affected ears and eyes in the offspring (cf. *Pop. Science Monthly*, New York, xi. 334). The important law of inheritance at corresponding periods was also set aside. Gangrene or inflammation commenced in both ears and both eyes soon after birth (pointing possibly to infection of some kind); the epileptic period commenced "perhaps two months or more after birth," while the loss of toes had occurred before birth. In no case, as Weismann points out, is the original mutilation of the nervous system ever transmitted. Even where an extirpated ganglion was never regenerated in the parent, the offspring always regained the part in an apparently perfect condition. On the whole the conflicting results ought to be as puzzling to those who may attribute them to a universal tendency to inherit the exact condition of parents as they are to those who, like myself, are sceptical as to the existence of such a law or tendency.

[60] The various results need to be fully and impartially recorded, and they should also be well tested and confirmed in proportion as they appear improbable and contrary to general experience. Professor Romanes has been carrying out the necessary experiments for some time past.

[61] Natural History Museum, central hall, third recess on the left.

[62] *Traité de l'Hérédité*, ii. 489; *Variation of Animals and Plants under Domestication*, i. 469. If injuries are inherited, why has the repeated rupture of the hymen produced no inherited effect?

[63] Compare the three cases of crooked fingers given in *Variation of Animals and Plants under Domestication*, ii. 55, 240.

[64] *Ibid.*, i. 460. Thus, where two brothers married two sisters all the seven children were perfect albinos, although none of the parents or their relatives were albinos. In another case the nine children of two sound parents were all born blind (ii. 322).

[65] See pp. 179-182, *Evolution and Disease*, by J. Bland Sutton, to whom and to our mutual friend Dr. D. Thurston I am indebted for information on various points.

[66] *Variation of Animals and Plants under Domestication*, ii. 290; i. 454.

MISCELLANEOUS CONSIDERATIONS.

TRUE RELATION OF PARENTS AND OFFSPRING.

It is difficult to entirely free ourselves from the flattering and almost universal idea that parents are true originators or creators of copies of themselves. But the main truth, if not the whole truth, is that they are merely the transmitters of types of which they and their offspring are alike more or less similarly moulded resultants. A parent is a trustee. He transmits, not himself and his own modifications, but the stock, the type, the representative elements, of which he is a product and a custodian in one. It seems [120] probable that he has no more definite or "particulate" influence over the reproductive elements within him than a mother over the embryo or a vessel over its cargo. Parent and offspring are like successive copies of books printed from the same "type." A battered letter in the "type" will display its effects in both earlier and later copies alike, but a purely extraneous or acquired flaw in the first copy is not necessarily repeated in subsequent copies. Unlike printer's type, however, the material source of heredity is of a fluctuating nature, consisting of competing elements derived from two parents and from innumerable ancestors.

Galton compares parent and child to successive pendants on the same chain. Weismann likens them to successive offshoots thrown up by a long underground root or sucker. Such comparisons indicate the improbability of acquired modifications being transmitted to offspring. [121]

That parts are developed in offspring independently of those parts in parents is clear. Mutilated parents transmit parts which they do not possess. The offspring of young parents cannot inherit the later stages of life from parents who have not passed through them. Cases of remote reversion or atavism show that ancestral peculiarities can transmit themselves in a latent or undeveloped condition for hundreds or thousands of generations. Many obvious facts compelled Darwin to suppose that vast numbers of the reproductive gemmules in an individual are not thrown off by his own cells, but are the self-multiplying progeny of ancestral gemmules. Galton restricts the production of gemmules by the personal structure to a few exceptional cases, and would evidently like to dispense with pangenesis altogether, if he could only be sure that acquired characters are never inherited. Weismann entirely rejects pangenesis and the inheritance of [122] acquired characters. This enables him to explain heredity by his theory of the "Continuity of the Germ-plasm." [67] Parent and offspring are alike successive products or offshoots of this persistent germ-substance, which obviously would not be correspondingly affected by modifications of parts in parents, and so would render the transmission of acquired characters impossible.

[123]

INVERSE INHERITANCE.

Mr. Galton contends that the reproductive elements become sterile when used in forming and maintaining the individual, and that only a small proportion of them are so used. [68] He holds that the next generation will be formed entirely, or almost entirely, from the residue of undeveloped germs, which, not having been employed in the structure and work of the individual, have been free to multiply and form the reproductive elements whence future individuals are derived. Hence the singular inferiority not infrequently displayed by the children of men of extraordinary genius, especially where the ancestry has been only of a mediocre ability. The valuable germs have been used up in the individual, and rendered sterile in the structure of his person. Hence, too, the "strong tendency to deterioration in the transmission of [124] every exceptionally gifted race." Mr. Galton's hypothesis "explains the fact of certain diseases skipping one or more generations," and it "agrees singularly well with many classes of fact;" and it is strongly opposed to the theory of use-inheritance. The elements which are used die almost universally without germ progeny: the germs which are *not* used are the great source of posterity. Hence, when the germs or gemmules which achieve development are either better or worse than the residue, the qualities transmitted to offspring will be of an inverse character. If brain-work attracts, develops *and sterilizes* the best gemmules, the ultimate effect of education on the intellect of posterity may differ from its immediate effect.

EARLY ORIGIN OF THE OVA.

As the ova are formed at as early a period as the rest of the maternal structure, Galton [125] notices that it seems improbable that they would be correspondingly affected by subsequent modifications of parental structure. Of course it is not certain that this is a valid argument. We know that the paternal half of the reproductive elements does not enter the ovum till a comparatively late stage in its history, and it is quite possible that maternal elements or gemmules may also enter the ovum from without. If reproductive elements were confined to one special part or organ, we should be unable to explain the reproduction of lost limbs in salamanders, and the persistent effect of intercrossing on subsequent issue by the same mother, and the propagation of plants from shoots, or of the begonia from minute fragments of leaves, or the development of small pieces of water-worms into complete animals.

[126]

MARKED EFFECTS OF USE AND DISUSE ON THE INDIVIDUAL.

These are, to some extent, an argument against the cumulative inheritance of such effects. When a nerve atrophies from disuse, or a duct shrivels, or bone is absorbed, or a muscle becomes small or flabby, it proves, so far, that the average effect of use through enormous ages is *not* transmitted. When the fibula of a dog's leg thickens by 400 per cent. to a size "equal to or greater than" that of the removed tibia which previously did the work, [69] it shows that in spite of disuse for countless generations, the "almost filiform" bone has retained a potentiality of development which is fully equal to that possessed by the larger one which has been constantly used. When, after being reared on the ailanthus, the caterpillars of the *Bombyx hesperus* [127] die of hunger rather than return to their natural food, the inherited effect of ancestral habit does not seem to be particularly strong. Neither is there any strongly-inherited effect of long-continued ancestral wildness in many animals which are easily tamed.

WOULD NATURAL SELECTION FAVOUR USE-INHERITANCE?

If use-inheritance is really one of the factors of evolution, it is certainly a subordinate one, and an utterly helpless one, whenever it comes into conflict with the great ruling principle of Selection. Would this dominant cause of evolution have favoured a tendency to use-inheritance if such had appeared, or would it have discouraged and destroyed it? We have already seen that use-inheritance is unnecessary, since natural selection will be far more effective in bringing [128] about advantageous modifications; and if it can be shown that use-inheritance would often be an evil, it then becomes probable that on the whole natural selection would more strongly discourage and eliminate it as a hostile factor than it might occasionally favour such a tendency as a totally unnecessary aid.

USE-INHERITANCE AN EVIL.

Use-inheritance would crudely and indiscriminately proportion parts to actual work done—or rather to the varying *nourishment and growth* resulting from a multiplicity of causes—and this in its various details would often conflict most seriously with the real necessities of the case, such as occasional passive strength, or appropriate shape, lightness and general adaptation. If its accumulated effects were not corrected by natural or sexual selection, horns and antlers would [129] disappear in favour of enlarged hoofs. The elephant's tusks would become smaller than its teeth. Men would have callosities for sitting on, like certain monkeys, and huge corns or hoofs for walking on. Bones would often be modified disastrously. Thus the condyle of the human jaw would become larger than the body of the jaw, because as the fulcrum of the lever it receives more pressure. Some organs (like the heart, which is always at work) would become inconveniently or unnecessarily large. Other absolutely indispensable organs, which are comparatively passive or are very seldom used, would dwindle until their weakness caused the ruin of the individual or the extinction of the species. In eliminating various evil results of use-inheritance, natural selection would be eliminating use-inheritance itself. The displacement of Lamarck's theory by Darwin's shows that the effects of use-inheritance often differ from those [130] required by natural selection; and it is clear that the latter factor must at least have reduced use-inheritance to the very minor position of comparative feebleness and harmlessness assigned to it by Darwin.

Use-inheritance would be ruinous through causing unequal variation in co-operative parts—of which Mr. Spencer may accept his own instances of the jaws and teeth, and the cave-crab's lost eyes and persistent eye-stalks, as typical examples. That the variation would be unequal seems almost self-evident from the varying rapidity and extent of the effects of use and disuse on different tissues and on different parts of the general structure. The optic nerve may atrophy in a few months from disuse consequent on the loss of the eye. Some of the bones of the rudimentary hind legs of the whale are still in existence after disuse for an enormous period. Evidently use-inheritance could not equally modify the turtle [131] and its shell, or the brain and its skull; and in minor matters there would be the same incongruity of effect. Thus, if the molar teeth lengthened from extra use the incisors could not meet. Unequal and indiscriminate variation would throw the machinery of the organism out of gear in innumerable ways.

Use-inheritance would perpetuate various evils. We are taught, for instance, that it perpetuates short-sight, inferior senses, epilepsy, insanity, nervous disorders, and so forth. It would apparently transmit the evil effects of over-exertion, disuse, hardship, exposure, disease and accident, as well as the defects of age or immaturity.

Would it not be better on the whole if each individual took a fresh start as far as possible on the advantageous typical lines laid down by natural selection? Through the long stages of evolution from primæval protoplasm upwards, such species as were least affected by use-inheritance [132] would be most free to develop necessary but seldom-used organs, protective coverings such as shells or skulls, and natural weapons, defences, ornaments, special adaptations, and so forth; and this would be an advantage—for survival would obviously depend on the *importance* of a structure or faculty in deciding the struggle for existence and reproduction, and not on the total amount of its using or nourishment. If natural selection had on the whole favoured this officious ally and frequent enemy, surely we should find better evidence of its existence.

Without laying undue stress upon the evil effects of use-inheritance, a careful examination of them in detail may at least serve to counter-balance the optimistic *a priori* arguments for belief in that plausible but unproven factor of evolution.

The benefits derivable from use-inheritance are largely illusory. The effects of *use*, indeed, are [133] generally beneficial up to a certain point; for natural selection has sanctioned or evolved organs which possess the property or potentiality of developing to the right extent under the stimulus of use or nourishment. But use-*inheritance* would cumulatively alter this individual adaptability, and would tend to fix the size of organs by the average amount of ancestral use or disuse rather than by the actual requirements of the individual. Of course under changed conditions involving increased or lessened use of parts it might become advantageous; but even here it may prove a decided hindrance to adaptive evolution in some respects as well as an unnecessary aid in others. Thus in the case of animals becoming heavier, or walking more, it would *lengthen* the legs although natural selection might require them to be shortened. In the Aylesbury duck and the Call duck, if use-inheritance has increased the dimensions of the bones and tendons of the leg, [134] natural selection has had to counteract this increase so far as length is concerned, and to effect 8 per cent. of shortening besides. If use-inheritance thickens bones without proportionally lengthening them, it would hinder rather than help the evolution of such structures as the long light wings of birds, or the long legs and neck of the giraffe or crane.

VARIED EFFECTS OF USE AND DISUSE.

The changes which we somewhat roughly and empirically group together as the effects of "use and disuse" are of widely diverse character. Thus bone, as the physiological fact, thickens under *alternations* of pressure (and the consequent increased flow of nourishment), but atrophies under a steadily continued pressure; so that if the use of a bone involved continuous pressure, the effect of such use would be a partial or total absorption of that bone. Darwin shows that [135] bone lengthens as well as thickens from carrying a greater weight, while tension (as seen in sailors' arms, which are used in pulling) appears to have an equally marked effect in shortening bones (*Descent of Man*, p. 32). Thus different kinds of use may produce opposite results. The cumulative inheritance of such effects would often be mischievous. The limbs of the sloth and the prehensile tail of the spider monkey would continually grow shorter, while the legs of the evolving elephant or rhinoceros might lengthen to an undesirable extent. Such cumulative tendencies of use-inheritance, if they exist, are obviously well kept under by natural selection.

Although the ultimate effect of use is generally growth or enlargement through increased flow of blood, the first effect usually is a loss of substance, and a consequent diminution of size and strength. When the loss exceeds the growth, use will diminish or deteriorate the part used, while disuse [136] would enlarge or perfect it. Teeth, claws, nails, skin, hair, hoofs, feathers, &c., may thus be worn away faster than they can renew themselves. But this wearing away usually stimulates the repairing process, and so increases the rate of growth; that is, it will increase the size produced, if not the size retained. Which effect of use does use-inheritance transmit in such cases—the increased rate of growth, or the dilapidation of the worn-out parts? We can hardly suppose that both these effects of use will be inherited. Would shaving destroy the beard in time or strengthen it? Will the continued shearing of sheep increase or lessen the growth of wool? What will be the ultimate effect of plucking geese's quills, and of the eider duck's abstraction of the down from her breast? If the mutilated parts grow stronger or more abundantly, why were the motmot's feathers alleged to be narrowed by the inherited effects of ancestral nibbling? [137]

The "use" or "work" or "function" of muscles, nerves, bones, teeth, skin, tendon, glands, ducts, eyes, blood corpuscles, cilia, and the other constituents of the organism, is as widely different as the various parts are from each other, and the effects of their use or disuse are equally varied and complicated.

USE-INHERITANCE IMPLIES PANGENESIS.

How could the transmission of these varied effects to offspring be accounted for? Is it possible to believe, with Mr. Spencer, that the effects of use and disuse on the parts of the personal structure are simultaneously registered in corresponding impressions on the seminal germs? Must we not feel, with Darwin apparently, [70] that the *only* intelligible explanation of use-inheritance [138] is the hypothesis of Pangenesis, according to which each modified cell, or physiological unit, throws off similarly-modified gemmules or parts of itself, which ultimately reproduce the change in offspring? If we reject pangenesis, it becomes difficult to see how use-inheritance can be possible.

PANGENESIS IMPROBABLE.

The more important and best-known phenomena of heredity do not require any such hypothesis, and leading facts (such as atavism, transmission of lost parts, and the general non-transmission of acquired characters) are so adverse to it that Darwin has to concede that many of the reproductive gemmules are atavistic, and that by continuous self-multiplication they may preserve a practical "continuity of germ-substance," as Weismann would term it. The idea that the [139] relationship of offspring to parent is one of direct descent is, as Galton tells us, "wholly untenable"; and the only reason he admits some supplementary traces of pangenesis into his "Theory of Heredity," [71] is that he may thus account for the more or less questionable cases of the transmission of acquired characters. But there appears to be no necessity even for this concession. We ought therefore to dispense with the useless and gratuitous hypothesis that cells multiply by throwing off minute self-multiplying gemmules, as well as by the well-known method of self-division. If pangenesis occurs, the transmission of acquired characters ought to be a prominent fact. The size, strength, health and other good or evil qualities of the cells could hardly fail to exercise a marked and corresponding effect upon the size and quality of the reproductive gemmules thrown off by those cells. The direct evidence tends to [140] show that these free gemmules do not exist. Transfusion of blood has failed to affect inheritance in the slightest degree. Pangenesis, with its attraction of gemmules from all parts of the body into the germ-cells, and the free circulation of gemmules in the offspring till they hit upon or are attracted by the particular cell or cells, with which alone they can readily unite, seems a less feasible theory and less in conformity with the whole of the facts than an hypothesis of germ-continuity which supposes that the development of the germ-plasm and of the successive self-dividing cells of the body proceeds from within. Darwin's keen analogy of the fertilization of plants by pollen renders development from without conceivable, but as there are no insects to convey gemmules to their destination, each kind of gemmule would have to be exceedingly numerous and easily attracted from amongst an inconceivable number of other gemmules. Arguments against pangenesis [141] can also be drawn from the case of neuter insects—a fact which seems to have escaped Darwin's notice, although he had seen how strongly that case was opposed to the doctrine which is the essential basis of the theory of pangenesis.

SPENCER'S EXPLANATION OF USE-INHERITANCE.

Mr. Spencer's explanation of the inheritance of the effects of use and disuse (p. 36) is that "while generating a modified *consensus* of functions and of structures, the activities are at the same time impressing this modified *consensus* on the sperm-cells and germ-cells whence future individuals are to be produced"—a proposition which reads more like metaphysics than science. Difficult to understand or believe in ordinary instances, such *consensus*-inheritance seems impossible in cases like that of the hive-bee. Can we suppose that the *consensus* of the activities of the working bee impresses [142] itself on the sperm-cells of the drones and on the germ-cells of the carefully secluded queen? Büchner thinks so, for he says: "Although the queens and drones do not now work, yet the capacities inherited from earlier times still remain to them, especially to the former, and are kept alive and fresh by the impressions constantly made upon them during life, and they are thus in a position to transmit them to posterity." Surely it is better to abandon a cherished theory than to be compelled to defend it by explanations which are as inconsistent as they are inadequate. New capacities are developed as well as old ones kept fresh. The massacre or expulsion of the drones would have to impress itself on the germ-cells of an onlooking queen, and the imprisonment of the queen on the sperm-cells of the drones—and in such a way, moreover, as to be afterwards developed into action in the neuters only. And use-inheritance all the while is being thoroughly [143] overpowered by impression-inheritance—by the full transmission of that which is merely seen in others! If such a law prevails, one may feel cold because an ancestor thought of the frosty Caucasus. None of this absurdity would arise if it were clearly seen that a parent is only a trustee—that transmission and development are perfectly distinct—that parental modifications are irrelevant to those transmitted to offspring.

FOOTNOTES:

[67] *Essays on Heredity*, p. 104. Weismann's theory is clear, simple and convenient, but incomplete; for, unlike Darwin's theory of pangenesis, it scarcely attempts any real explanation of the extremely complex potentialities possessed by the reproductive elements. Perhaps we might retain Darwin's self-multiplying gemmules without supposing them to be thrown off by the cells, which will no longer be credited with *two* modes of multiplication. These minute germs or gemmules may have been evolved by natural selection playing upon the sample germs that achieve development; and they may exist either separately, or (preferably but perhaps not invariably) in aggregates to form Weismann's germ-plasm.

[68] *Contemporary Review*, Dec., 1875, p. 88.

[69] *Variation of Animals and Plants under Domestication*, ii. 286.

[70] *Variation of Animals and Plants under Domestication*, ii. 388, 398, 367; *Life and Letters*, iii. 44.

[71] *Contemporary Review*, Dec., 1875, pp. 94, 95.

[144]

CONCLUSIONS.

USE-INHERITANCE DISCREDITED AS UNNECESSARY, UN-PROVEN, AND IMPROBABLE.

General experience teaches that acquired characters are not usually inherited; and investigation shows that the apparent exceptions to this great rule are probably fallacious. Even the alleged instances of use-inheritance culled by such great and judicious selectors as Darwin and Spencer break down upon examination; for they can be better explained without use-inheritance than with it. On the other hand, the adverse facts and considerations are almost strong enough to prove the actual non-existence of such a law or [145] tendency. There is no need to undertake the apparently impossible task of demonstrating an absolute negative. It will be enough to ask that the Lamarckian factor of use-inheritance shall be removed from the category of accredited factors of evolution to that of unnecessary and improbable hypotheses. The main explanation or source of the fallacy may be found in the fact that natural selection frequently imitates some of the more obvious effects of use and disuse.

MODERN RELIANCE ON USE-INHERITANCE MISPLACED.

Modern philanthropy — so far at least as it ever studies ultimate results — constantly relies on this ill-founded belief as its justification for ignoring the warnings of those who point out the ultimately disastrous results of a systematic defiance or reversal of the great law of natural selection. [146] This reliance finds strong support in Mr. Spencer's latest teachings, for he holds that the inheritance of the effects of use and disuse takes place universally, and that it is now "the chief factor" in the evolution of civilized man (pp. 35, 74, iv) — natural selection being quite inadequate for the work of progressive modification. Practically he abandons the hope of evolution by natural selection, and substitutes the ideal of a nation being "modified *en masse* by transmission of the effects" of its institutions and habits. Use-inheritance will "mould its members far more rapidly and comprehensively" than can be effected by the survival of the fittest alone.

But could we rely upon the aid of use-inheritance if it really were a universal law and not a mere simulation of one? Let us consider some of the features of this alleged factor of evolution, seeing that it is henceforth to be our principal means of securing the improvement of our species [147] and our continued adaptation to the changing conditions of a progressive civilization.

It is curiously uncertain and irregular in its action. It diminishes or abolishes some structures (such as jaws or eyes) without correspondingly diminishing or abolishing other equally disused and closely related parts (such as teeth, or eye-stalks). It thickens ducks' leg-bones while allowing them to shorten. It shortens the disused wing-bones of ducks and the leg-bones of rabbits while allowing them to thicken; and yet in other cases it greatly reduces the thickness of bones without shortening them. It transmits tameness most powerfully in an animal which usually cannot acquire it. It aids in webbing the feet of water-dogs, but fails to web the feet of the water-hen or to remove the web in the feet of upland geese. [72] It allows the disused fibula to retain a [148] potentiality of development fully equal to that possessed by the long-used tibia. It lengthens legs because they are used in supporting the body, and shortens arms because they are used in pulling. Whether it enlarges brain if used in one way and diminishes it if used in another, we cannot tell; but it must obviously deaden nervous sensibilities in some cases and intensify them in others. It enlarges hands long before they are used, and thickens soles long before the time for walking on them. At the same time, as if by an oversight, it so delays its transmission of the habit of walking on these thickened soles, that the gradual and tedious acquisition of the non-transmitted habit costs the infant much time and trouble and often some pain and danger. Yet where aided by natural selection, as with chickens and foals, it transmits the habit in wonderful perfection and at a remarkably early date. It transmits new paces in horses in [149] a single generation, but fails to perpetuate the songs of birds. It modifies offspring like parents, and yet allows the formation of two reproductive types in plants, and of two or more types widely different from the parents in some of the higher insects. It is said to be indispensable for the co-ordinated development of man and the giraffe and the elk, but appears to be unnecessary for the evolution and the maintenance of wonderful structures and habits and instincts in a thousand species of ants and bees and termites. It is the only possible means of complex evolution and adaptation of co-operative parts, and yet in Mr. Spencer's most representative case it renders such important parts as teeth and jaws unsuited for each other, and is said to ruin the teeth by the consequent overcrowding and decay. It survives amidst a general "lack of recognised evidence," and only seems to act usefully and healthily and regularly in quarters where it can least easily be [150] distinguished from other more powerful and demonstrable factors of evolution. So little does it care to display its powers where they would be easily verifiable as well as useful that practical breeders ignore it. So slight is its independent power that it seems to allow natural selection or sexual selection or artificial selection to modify organisms in sheer defiance of its utmost opposition, just as readily as they modify organisms in other directions with its utmost help. If it partially perpetuates and extends the pecked-out indentations in the motmot's tail feathers, it on the other hand fails to transmit the slightest trace of mutilation in an almost infinite number of ordinary cases, and even where the mutilation is repeated for a hundred generations; and it apparently repairs rather than transmits the ordinary and oft-repeated losses caused by plucking hair, down and feathers, and the wear and tear of claws, teeth, hoofs and skin.

[151]

It is often mischievous as well as anomalous in its action. Under civilization with its division of labour, the various functions of mind and body are very unequally exercised. There is overwork or misuse of one part and disuse and neglect of others, leading to the partial breakdown or degeneration of various organs and to general deterioration of health through disturbed balance of the constitution. The brain, or rather particular parts of it, are often overstimulated, while the body is neglected. In many ways education and civilization foster nervousness and weakness, and undermine the rude natural health and spirits of the human animal. Alcohol, tobacco, tea, coffee, extra brain work, late hours, dissipation, overwork, indoor life, division of labour, preservation of the weak, and many other causes, all help to injure the modern constitution; so that the prospect of cumulative intensification of these evils by the additional influence of use-inheritance is not an encouraging one. It is true [152] that modern progress and prosperity are improving the people in various respects by their direct action; but if use-inheritance has any share in effecting this improvement it must also transmit increased wants and more luxurious habits, together with such evils as have already been referred to. As depicted by its defenders, use-inheritance transmits evils far more powerfully and promptly than benefits. It transmits insanity and shattered nerves rather than the healthy brain which preceded the breakdown. It perpetuates, and cumulatively intensifies, a deterioration in the senses of civilized men, but it fails to perpetuate the rank vigour of various plants when too well nourished, or the flourishing condition of various animals when too fat or when tamed. It already transmits the short-sight caused by so modern an art as watchmaking, but so fails to transmit the long-practised art of seeing (as it does of walking and talking) that vision is worse than useless to a man until he gradually acquires the [153] necessary but non-transmitted associations of sensation and idea by his own experience. In a well-known case, a blind man on gaining his sight by an operation said that "all objects seemed to touch his eyes, as what he felt did his skin"—so little had the universal experience of countless ages impressed itself on his faculties. Under normal healthy conditions use-inheritance is so slow in its action that "several generations" must elapse before it produces any appreciable effect, and then that effect is only precisely what selection might be expected to bring about without its aid. Strong for evil and slow for good, it can convey epilepsy promptly in guinea pigs, but transmits the acquirements of genius so poorly that our best student of the heredity of genius has to account for the frequent and remarkable deterioration of the offspring by a theory which is strongly hostile to use-inheritance. It would tend to make organisms unworkable by the excessive differences in its rate [154] and manner of action on co-operative parts, and by adapting these parts to the total amount of nourishment received rather than to occasional necessity or actual usefulness. It would tend to stereotype habits and convert reason into instinct.

How then can we rely upon use-inheritance for the improvement of the race? Even if it is not a sheer delusion, it may be more detrimental as a positive evil than it is advantageous as an unnecessary benefit; and as a normal modifying agent it is miserably weak and untrustworthy in comparison with the powerful selective influences by which nature and society continually and inevitably affect the species for good or for evil. The effects of use and disuse—rightly directed by education in its widest sense—must of course be called in to secure the highly essential but nevertheless *superficial, limited, and partly deceptive* improvement of individuals and of social manners and methods; but as this artificial development of already existing [155] potentialities does not directly or readily tend to become congenital, it is evident that some considerable amount of natural or artificial selection of the more favourably varying individuals will still be the only means of securing the race against the constant tendency to degeneration which would ultimately swallow up all the advantages of civilization. The selective influences by which our present high level has been reached and maintained may well be modified, but they must not be abandoned or reversed in the rash expectation that State education, or State feeding of children, or State housing of the poor, or any amount of State socialism or public or private philanthropy, will prove permanently satisfactory substitutes. If ruinous deterioration and other more immediate evils, are to be avoided, the race must still be to the swift and the battle to the strong. The healthy Individualism so earnestly championed by Mr. Spencer must be allowed free play. Open competition, as Darwin teaches, with [156] its survival and multiplication of the fittest, must be allowed to decide the battle of life independently of a foolish benevolence that prefers the elaborate cultivation and multiplication of weeds to the growth of corn and roses. We are trustees for the countless generations of the future. If we are wise we shall trust to the great ruling truths that we assuredly know, rather than to the seductive claims of an alleged factor of evolution for which no satisfactory evidence can be produced.

THE END.
RICHARD CLAY AND SONS, LIMITED, LONDON AND BUNGAY.

FOOTNOTES:

[72] Professor Romanes had casts made of the feet of upland geese, and could not detect any diminution as compared with the web of other geese in relation to the toes.

www.ingramcontent.com/pod-product-compliance
Lightning Source LLC
Chambersburg PA
CBHW030451220526
45464CB00006B/2495